The Myth of the OPEC Cartel

The Role of Saudi Arabia

The Myth of the OPEC Cartel

The Role of Saudi Arabia

Ali D. Johany
Dean of the College of Industrial Management
University of Petroleum and Minerals
Dhahran, Saudi Arabia

UNIVERSITY OF PETROLEUM AND MINERALS
Dhahran, Saudi Arabia

and

JOHN WILEY & SONS
Chichester · New York · Brisbane · Toronto · Singapore

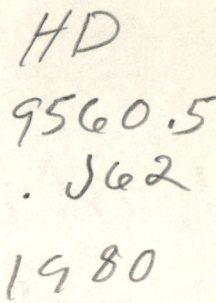

HD
9560.5
.J62
1980

Copyright © 1980 by John Wiley & Sons Ltd.

Corrected reprint March 1982

All rights reserved.

No part of this book may be reproduced by any means, nor transmitted, nor translated into a machine language without the written permission of the publisher.

British Library Cataloguing in Publication Data:

Johany, Ali D.
   The myth of the OPEC cartel.
   1. Organisation of Petroleum Exporting Countries—History
   2. Petroleum—Prices—History
   3. Petroleum industry and trade—Saudi Arabia
   —History
   I. Title
   338.2'3   HD9560.1.066   80–40959

ISBN 0 471 27864 5

Typeset by Preface Ltd, Salisbury, Wilts. and printed by The Pitman Press, Bath, Avon.

Table of Contents

	Page
Preface	vii
List of Tables	ix
List of Figures	xi

1 Introduction 1
   A Personal Comment 2

2 The Evolution of OPEC 4
   2.1 The Early Stages 4
   2.2 The Formation of OAPEC 6
   2.3 The Changing Role of the 'Seven Sisters' 7
   2.4 Libya Leads OPEC 9
   2.5 The Last Hurrah: The Quadrupling of Prices 13

3 The Stability of Cartels and Price Leadership 17
   3.1 Preliminary Remarks 17
   3.2 The Difficult Task of Colluding 19
   3.3 Collusion Within OPEC 20
   3.4 Why Cartels? 21
   3.5 A Theory of Cartels 23
   3.6 OPEC and Collusion 25
   3.7 OPEC as a Joint Sales Agency 26
   3.8 Price Leadership by a Dominant Firm 28
   3.9 The Formal Model 29

4 The Economics of Exhaustible Resources 33
   4.1 The Formal Model 33
   4.2 The Threat to Property Rights 37

5 OPEC is not a Cartel: an Alternative Explanation 40
5.1 The Oil Market: 1947–1970 40
5.2 The Oil Market: 1970–1973 40
5.3 Cartelization or Alteration of Property Rights . . . 43
5.4 But Does Not Saudi Arabia Dominate OPEC? 49
5.5 An Alternative Hypothesis: The Companies As Tax Collectors . 52

6 The Iranian Revolution and the Oil Price Explosion 54
6.1 Temporary and Permanent Effects 54
6.2 The Evolution of Prices, January 1978–June 1979 55
6.3 The Explosion of Prices, November 1979–February 1980. . . 56
6.4 Blaming the Oil Companies 58

7 Saudi Economic Choices 60
Introduction 60
7.1 Absorptive Capacity 61
7.2 The Saudi Realities 63
7.3 The Risks and Rewards of Foreign Investment 66
7.4 OPEC and Saudi Arabia 71
A *Word* on the Appendices 72
Appendix A The Nature of Inflation in Saudi Arabia 73
Appendix B The Price of Land in Saudi Arabia 75
Appendix C Private Saudi Investment Abroad 77
Appendix D Not For Rent Please: The Consequences of Rent
           Controls 79

8 Saudi Institutional Processes and Economic Modernization 81
Introduction 81
8.1 Decision Making 82
8.2 The Ministry of Finance and National Economy: A Government
    Within Government 83
8.3 Saudi Economic Development 84
8.4 Saudi Economic Issues in the 1980s 89
    (A) Income Distribution 89
    (B) The State of the Oil Fields 91
    (C) Water 93

9 The World Oil Market and the Role of Saudi Arabia: A Summary . 95

Bibliography 100

Index 106

Preface

It is widely believed that the current price of crude oil reflects the monopoly power that the 'OPEC Cartel' has been able to exercise. In this book, however, it will be argued that the presence or absence of OPEC has nothing to do with the prevailing world oil price. This conclusion may shock many readers who were led to believe by journalists and professional economists alike that the two words 'OPEC Cartel' were a single, inseparable expression.

In fact when this book was started, it did not occur to me for a single moment that OPEC might not be a cartel. The original purpose of this book was to try to understand why the 'OPEC Cartel' was so different from other cartels in its ability to prosper and survive; after all, economists have told us repeatedly that cartels have short and unhappy lives.

It turned out that economists are in fact right in believing that cartels are inherently unstable if they are able to earn monopoly profits for their members. If OPEC was a cartel, it would have been no exception.

Then, how could we explain the magnitude and suddenness of the world oil price rise which occurred toward the end of 1973?

This book is largely an attempt to answer this question. But the brief response to the query is this: since the 1950s the Western oil companies were never one hundred per cent certain that their property rights on crude oil deposits would not be one day in jeopardy; consequently, they extracted more oil during the 1950s and 1960s than they otherwise would have, and the result was lower-than-otherwise prices. A chain of events in late 1973, culminating in the historic six Gulf producers meeting in Kuwait on October 16, had radically changed the world oil market. It was decided during that meeting that henceforth each country would set its own oil price with no consultation or advance negotiation with the foreign oil companies as had been hitherto done; that amounted to a **de facto** nationalization of crude oil deposits. The oil-producing countries, on the other hand, do not face ownership uncertainty, and thus have a longer time horizon, which would dictate smaller amounts of output; the result is higher-than-otherwise oil prices.

Saudi Arabia plays a major role in the world oil market, not only because of the sheer size of its proven oil reserve, but also because of its willingness to increase its oil output whenever the price of oil rises. Thus, two chapters are entirely devoted to a discussion of the Saudi economy and to the process of decision-making in the Kingdom of Saudi Arabia.

This book may seem technical and thus hard to understand. In reality it is not. Although it contains many mathematical symbols and complicated diagrams, all of its findings are also verbally explained.

For their helpful comments and useful suggestions on different parts of the manuscript I must thank Professors H. E. Frech III, Robert Deacon, J. Wilson Mixon and, above all, Professor Walter J. Mead.

Dhahran, Saudi Arabia   ALI D. JOHANY
May, 1980.

List of Tables

		Page
Table 5.1	Free world crude oil output in millions of barrels per day	42
Table 5.2	Percentage change in gross domestic product of the OECD countries	42
Table 5.3	International crude oil production for major petroleum exporting countries	46
Table 5.4	OAPEC members of OPEC production levels during the 3-month 'embargo' period October through December 1973	48
Table 5.5	OPEC country output and market shares, classified by expanding and contracting countries	50
Table 5.6	OPEC country output and market shares, classified by 'saver countries' and 'spender countries'	51
Table 7.1	Gross domestic product by economic activity at current prices	64
Table 7.2	The price of Saudi Arabian Crude	66
Table 7.3	Oil revenue by source (million US dollars)	67
Table 7.4	Government actual revenue and expenditure (million riyals)	67
Table 8.1	Real GDP (annual growth rates)	85
Table 8.2	Annual growth rates of money supply and real supplies of goods and services	86
Table 8.3	Project budget expenditure (million riyals)	87
Table 8.4	Road network in the kingdom (cumulative length in km)	88

List of Figures

		Page
Figure 3.1	The industry's incentives to collude	22
Figure 3.2	The incentive for individual firms to depart from the collusive solution	22
Figure 3.3	Industry demand $D(P)$ and small firms' supply $S(P)$	30
Figure 3.4	Choice of price P_d and output Q_d for the dominant firm	31
Figure 3.5	Industry equilibrium with a dominant firm producing $Q_d = \bar{Q} - Q_s$	31
Figure 4.1	Price changes over time in the case of an exhaustible resource	36
Figure 4.2	The behavior of price over time in the case of constant costs	36
Figure 4.3	The influence of increasing the discount rate on prices over time and on the date of exhaustion	38
Figure 5.1	The oil market, 1947–1970	41
Figure 5.2	The oil market, 1973–1974	48
Figure 7.1	Allocation of investment funds	68
Figure 7.2	The allocation of investment funds under conditions of risk to property rights (r_f and r_d are riskless rates of return)	70

CHAPTER 1
Introduction

The world price of crude oil is determined in the same way as the price of any other commodity—by demand, production costs, and the extent of noncompetitive behaviour. What makes the oil market interesting from a theoretical point of view is the claim that the current price is forty to fifty times the current marginal cost of production.

It is widely believed that the current price of crude is arbitrary and reflects the monopoly power of the Organization of Petroleum Exporting Countries (OPEC) rather than the long-run supply costs.

The purpose of this study is to investigate this claim and to answer the question: 'Is OPEC a cartel?' If the answer is 'yes', then what makes this cartel so robust? More specifically, how are the members of OPEC able to agree on their oil price levels, and how do they share their 'monopoly profits'?

If OPEC is not a cartel, then we must explain the apparent huge price increase that occurred between January 1973 and January 1974.

The problem is a very important one for theoretical and policy reasons. If, in fact, OPEC is a cartel, then its achievements are truly remarkable because, on theoretical grounds, one would expect the success of a cartel to lead to its own destruction (see Chapter 3). The greater the monopoly profits that any cartel is able to reap, the greater the incentives for each individual member to depart from the cartel solution.

The policy implications to the oil-consuming countries, if OPEC is not a cartel, are equally important. One of the strongest arguments that is frequently offered to defend the now defunct US oil price controls is the claim that the world oil price is a monopoly price and, therefore, US domestic oil producers should not be allowed to receive monopoly prices for their oil. It is thought that if the oil was deregulated it would have been in the American oil industry's interest to cooperate rather than compete with OPEC to keep the world oil prices at high levels.

Our purpose is to understand OPEC in general and Saudi Arabia's behaviour in particular as the world's largest oil exporter. The main concern is to understand the forces that determine the supply of oil. But to understand these forces, we have to describe the institutional factors that are peculiar to the oil market. Thus Chapter 2 gives a brief history of the evolution of OPEC: why it was created and how.

To specialists, this chapter will add very little to what is already known. To non-specialists, on the other hand, this chapter is mandatory reading. It gives a brief, but thorough, treatment of the circumstances that surrounded the creation of OPEC.

From this chapter, the reader will learn OPEC's historical importance in enabling the oil-producing countries to become the 'real' owners of the oil deposits in their own country. He or she will also learn that the special political environment which the Arab–Israeli conflict had created in fact played a large role in accelerating the transfer of oil ownership from the companies to their host countries.

Chapter 3 starts with preliminary remarks concerning market structures and the circumstances that surround the final emergence of monopoly or competition for any particular product.

Sections 3.4 and 3.5 explain models of economic collusion that are chiefly based on the works of the University of Chicago's George Stigler and Gary Becker. Collusion is viewed here in a cost–benefit framework. The final result is stated in a single functional form. Then we try to see how applicable the model is to OPEC. The nature of price-leadership by a dominant firm is explained in sections 3.8 and 3.9.

In Chapter 4, the special nature of the economics of exhaustible resources is reviewed. It also contains a simple probabilistic model that explains how uncertainty of property rights radically changes the discount rates, which in turn have great influence on the size of oil output in each period of time.

In Chapter 5, the conclusion that OPEC is in fact not a cartel is reached. In this chapter the movements of the world crude oil prices from 1947 to 1973 are traced. The role of Saudi Arabia in formulating the 'OPEC Price' is explained in section 5.4.

What happened in the oil market since the beginning of the Iranian revolution is discussed in Chapter 6.

The factors that influence Saudi decisions are briefly explained in Chapters 7 and 8.

Finally, Chapter 9 summarizes the main findings of the entire book.

A PERSONAL COMMENT

Because I happen to be from Saudi Arabia it is inevitable, I think, that I will be accused of being an apologist for OPEC. It is frequently said that economists, like other social scientists, are necessarily ideological.

If I may borrow the words of MIT's Robert Solow, I do not deny that economists, like everyone else—social scientists or not—have interests and ideological commitments. Choosing OPEC as my topic of research is obviously a reflection of my values and interests. However, I would like to assure the reader that my own conclusion was as much a surprise to me as it might well be to him or her.

Professor Robert Solow, one of the world's most eminent economists, expressed very eloquently how mistaken one could be if he thinks the conclusions of social scientists should be dismissed simply because social sciences are not 'value-free':

> The whole discussion of value-free social science suffers from being conducted in qualitative instead of quantitative terms. Many people seem to have rushed from the claim that no social science can be perfectly value-free to the conclusion that therefore anything goes. It is as if we were to discover that it is impossible to render any operating room perfectly sterile and conclude therefore one might as well do surgery in a sewer (*The Public Interest*, Fall 1970: 94–107).

CHAPTER 2
The Evolution of OPEC

2.1 THE EARLY STAGES

The Western oil companies began looking for oil in Latin America and the Middle East in the early 1930s and 1940s. At that time, most of these countries were either completely controlled by the Western powers or totally unfamiliar with petroleum politics, engineering and—most significantly—economics. The oil companies were in a position, therefore, to determine the terms of the concessions that allowed them to explore for and produce oil.

The terms of the early concessions dictated that the producing countries received a 'royalty' which was a fixed nominal amount per unit of output (Arab League, 1959). The revenues of the producing countries varied with volume of output. To the producing countries the price of oil had no direct effect on the oil revenues. But not for long.

By 1949, Venezuela, which had the greatest degree of independence from the West, and possessed a more advanced technical knowledge, was able to get 50 per cent of nominal profits through income taxes. This profit-sharing formula, later to become known as the 50–50 arrangement, provided that the host government's 50 per cent share of profits include royalties, e.g., if nominal profits were 100 and royalties 12, the government share is 38. Nominal profit per unit of output was defined as posted price minus production costs—however defined (Mikdashi, 1972).

The notion of posted price had its origin in the US Texas Gulf, where buyers of crude made public the prices at which they intended to buy oil. Gradually the posted price became a proxy for the open market price and was used to determine payments to producing countries.

The important point here is that the cost of Venezuelan crude became much higher to the oil companies than the cost of oil in the Arabian Gulf area. The Venezuelan Government got concerned about the threat of the cheaper oil in the Middle East. And an official Venezuelan delegation paid a series of visits to the oil-producing countries in the Arabian Gulf, to explain the advantages of the 50–50 arrangement instituted in Venezuela.

By the end of 1950 Iraq, Kuwait, and Saudi Arabia had successfully negotiated the 50–50 arrangements (Mikdashi, 1972).

In Iran, the oil companies refused to budge, and that contributed to the political crisis of 1951 and the nationalization of the oil industry by the Iranian government. The political upheaval in Iran was a warning to all the oil companies, even though their property rights in Iran were restored in 1954, to give up all resistance to the 50–50 arrangement.

Thus the era of royalties was essentially replaced by the era of profit sharing. The companies paid royalties and taxes on the basis of the posted price even if the price at which oil could be sold in the marketplace was below the posted price, which has been the case since the early 1950s (Chandler, 1973).

Posted prices were a sort of internal accounting price that the oil companies used to transfer oil from the producing subsidiaries to the refining subsidiaries. A tax on profits meant that the producing countries had to know the receipts and outlays of the companies. In fact, the host countries had to take the companies' word. This led to the recognition on the part of the producers that they needed to cooperate in the collection and interpretation of information on oil costs and revenues.

As a UCLA economist will tell you, the oil producing countries were trying to reduce the costs of information. Contacts between the oil producers were made on a regular basis following the 1950 efforts by the Venezuelans and, in April 1959, the first Arab Petroleum Congress met in Cairo. Representatives from Iran and Venezuela attended as observers (Arab League, 1959). In the words of the founding father of OPEC (the Godfather, if you like), Perez Alfonzo of Venezuela, it was during this Cairo conference that 'the first seed of the creation of OPEC was sown' (Rouhani, 1971: 36).

Whoever gets the credit, the immediate cause for the creation of OPEC was the posted price reductions of the 1958–60 period. Why the reductions?

On the demand side, the US Government imposed mandatory import quotas (which meant only a specific quantity could be imported) on both crude oil and oil products in order to reduce 'dependence on foreign oil'. The quota was most likely to protect domestic producers, but claiming that the purpose of the quota is to make one's country self sufficient is, of course, as every lobbyist knows, a more persuasive argument. But Professor M. A. Adelman disputes this widely held view that the US import quota had significantly influenced the oil market. His contention is that the American States oil regulation had long before the Federal government 1959 action succeeded in curtailing oil imports (Adelman, 1972: Chapter 5). That being as it may, no one will deny that American official policy, whether state or federal, reduced American demand for foreign oil. On the supply side the entry of the so-called 'independents' and the State oil companies of consumer countries (i.e., French, Italian, and Japanese) had increased the supply of crude by a higher rate than was expected (Jacoby, 1974). The result of a fall in demand and an increase of supply was to reduce the posted price and thus the oil revenues of the producing countries.

From 10–14 September, 1960, the representatives of Iran, Iraq, Kuwait, Saudi Arabia, and Venezuela conferred in 'an atmosphere of crisis' in Baghdad and brought into being the Organization of Petroleum Exporting Countries (OPEC) as a permanent intergovernmental organization (Rouhani, 1971). Eight other countries subsequently joined OPEC: Qatar (1961), Indonesia and Libya (1962), Abu Dhabi—later to be a part of the United Arab Emirates (UAE)—(1967), Algeria (1969), Nigeria (1971), and Ecuador and Gabon (1973).

The objectives of OPEC were set forth as follows:

1. The principal aim of the organization shall be the coordination and unification of the petroleum policies of member countries and the determination of the best means for safeguarding their interests, individually and collectively.
2. The organization shall devise ways and means of ensuring the stabilization of prices ... [and] eliminating harmful and unnecessary fluctuations (OPEC, 1976).

It was partially successful; no reduction of nominal prices has occurred since OPEC was established, even though the downward pressure on real prices continued during the 1960s.

The negotiations of 1962–63 between the companies and OPEC changed the treatment of royalties from an income tax to an expense. The government revenue per barrel was calculated, starting from 1964, as follows: royalties + 10 per cent [of receipts over and above costs—royalties], in effect abolishing the original 50–50 arrangement in favour of producers and moving away from income taxes to per-barrel payments. The new agreement resulted in increasing the per-barrel revenues by four cents. The companies were granted 0.5¢ per barrel marketing allowance off posted price. Other allowances based on complicated formulas amounting to 8.5 per cent of posted price were also granted. In 1966–67 another round of negotiations provided that the allowable discounts be phased out through 1973 (Mikdashi, 1972).

From 1967 to 1970 the accomplishments of OPEC were thought of as minor, and few people paid any attention to its activities.

2.2 THE FORMATION OF OAPEC

Following the 1967 Middle East War and the occupation by Israel of substantial areas of Arab territories, the Arabs became convinced (rightly or wrongly) that the US Government foreign policy reflected its partiality in favor of Israel. The Arab world demanded from the Arab oil-producing countries some action to counter what they perceived to be the anti-Arab activities of the US Government.

The Arab countries that exported oil found themselves faced with difficult choices. If they ignored the Arab world demands, the anti-American sentiment might get out of hand and lead to further Arab radicalization and massive sabotage of oil installations. On the other hand, the governments of some oil-producing countries had established a special relation with the US and did not want to change that relation.

But whatever non-financial objective the Arabs wanted to achieve could not be effectuated through OPEC, which included non-Arab members with different interests. Accordingly, in January 1968, Saudi Arabia, Kuwait, and Libya signed an agreement in Beirut creating the Organization of Arab Petroleum Exporting Countries (OAPEC) (Mikdashi, 1972). Iraq and Algeria refused to join OAPEC at that time because they had political differences with the founding members of OAPEC. Things changed following the Libyan Revolution of September 1, 1969, which replaced the ultra-conservative monarchy with the ultra-nationalist government of Muammar Ghaddafi. By 1972 all of the Arab oil-exporting countries became members of OAPEC regardless of the importance of the oil sector to their economies.

After 1967 a short-lived, ill-conceived—it was opposed by Saudi Arabia—embargo on petroleum going to the US, West Germany and the UK failed to have any impact (because the US oil imports were only 300,000 barrels per day and because the oil companies increased their rate of output in Venezuela and Iran). The Arab oil-producing countries decided that the best way to use oil as 'a weapon' was to transfer some of the oil revenues to the so-called 'confrontation states' of Egypt, Jordan and Syria. In short, OAPEC had little influence on the world petroleum market until October 1973, when indirectly it contributed to the adoption of actions that changed the oil market as it had hitherto been known.

2.3 THE CHANGING ROLE OF THE 'SEVEN SISTERS'

Up to 1950, the production and distribution of world oil (at that time excluding the US market) was effectively controlled by Exxon, British Petroleum (BP), Royal Dutch-Shell, Gulf, Texaco, Standard of California, and Mobil—the 'Seven Sisters', as they came to be known. Four of the seven (Exxon, BP, Shell, and Gulf) produced 83 per cent of the world's crude oil (Adelman, 1972).

From 1928 to 1948 the price of oil was determined by a single base point system. The price of oil in the US Gulf of Mexico determined the price of oil anywhere in the world. Thus Europe, for example, would pay the Texas Gulf price plus the cost of freight from the Texas Gulf to Europe, even though the oil might have been produced in the

Arabian Gulf at a cost one-fifth that of Texas and a freight cost 40 per cent that from Texas to Europe (Adelman, 1972).

As a result of such a system, the oil companies realized huge rates of return on their international investment, and that attracted newcomers to the market. Unless there is some economic or institutional force that prevents entry, high rates of return always attract unwelcome company.

In the early 1950s competitive pressure began to arise in the market. The single base point was changed to a double base (Texas and the Arabian Gulf). Concern about the power of the major oil companies led the French, Italian, and Japanese governments to finance state-owned companies to engage in the production, purchasing, and marketing of crude oil and its products. In the 1960s, private oil companies (such as Occidental, Standard of Indiana, Continental, Phillips, and Marathon) gained valuable concessions in areas outside the traditional sphere of influence of the majors, namely, Nigeria and Libya (Jacoby, 1974).

The economic implications of entry into the world oil market by the smaller private American companies, the 'independents', or the 'newcomers' and the state-owned European and Japanese companies, were not only to increase oil supplies but also to *provide* the oil-producing countries with *alternative* agents to deal with. Prior to these entries the only buyers of crude oil were the Seven Sisters. They (the 'Seven Sisters') enjoyed what economists call 'monopsony' power—the market power of a single buyer. The decrease of the market power of the 'Seven Sisters' (i.e., the elimination of their monopsony power) increased the bargaining power of the oil-producing countries.

Those developments in the international oil market played an important role in the later success of the OPEC members in independently setting the rates of output and prices of their oil. Due to the emergence of competition and the constant threat of takeover by the host countries that led the oil companies to act as if there was 'no tomorrow', the real price of oil at the Arabian Gulf decreased by 65 per cent between 1947 and early 1970 (Adelman, 1972).

It is important to emphasize that, during the first ten years of OPEC's life, important *structural changes* took place in the international oil market, but quite independently of OPEC. 'In my view,' wrote Professor E. T. Penrose, a distinguished British oil economist in 1969, 'OPEC has never been as powerful in the international [oil] industry generally as were the major companies in their heyday. But this heyday did not last long; it began to fade in the early fifties, although the fading did not become obvious until the end of that decade' (Penrose, 1969).

The important structural change in the international oil market in the 1960s was the increase of production of crude oil due to the entry and expansion of companies new to the industry. The 'major' oil companies could not sell their crude at the existing posted prices, but would not lower posted prices fearing political pressure in the host countries.

Professor Penrose, in analysing the oil market in 1969, very accurately predicted the outcome.

> Increasing monopoly revenues are not consistent with decreasing monopoly power. Either the governments [of exporting countries] or the companies will give way, and if OPEC succeeds in maintaining the position of the governments, they may well find themselves with a greater responsibility for their own industry than they either expect or want (Penrose, 1969: 433).

In fact, that is what happened, although I am not sure that OPEC, as an organization, was the cause.

2.4 LIBYA LEADS OPEC

One would expect that the producing-countries' first successful attempt toward output restriction would be made in Saudi Arabia, or Iran, or Kuwait, or Iraq, or even Venezuela—the established leaders and the founding members of OPEC, not in Libya, at the edge of the Arab World. But so it was. Nevertheless, Libya, by itself a minor producer, could not accomplish much.

By the end of 1969 Algeria was very unhappy with the 1965 Franco–Algerian agreement. The government was concerned that the rate of investment of the companies was small and the discovery of more reserves as a result was hindered. The entry of Getty Petroleum Company, an independent American oil company, in October increased the bargaining power of the Algerian Government. Prior to that, in May of 1969, Iran forced the companies to increase their payments by a substantial amount by threatening to take over 50 per cent of the developed fields. After the December 1969 meeting, OPEC members offered strong support to both Algeria and Libya, which were in conflict with the companies (Penrose, 1975).

In January 1970 negotiations started between Libya and the companies. The Libyan government wanted to increase royalties and the income tax rate. Pressure was concentrated on Occidental (Oxy) and Exxon. To increase the pressure on the companies, in June Ghaddafi ordered cutbacks in output for all the twenty operating companies. Oxy was forced to cut output by more than 50 per cent.

Exxon could not possibly yield due to its large holdings in Iran, Iraq, and Saudi Arabia. Should Exxon accept the Libyan demands, then those countries would insist on similar terms. Libya received strong support from Algeria and Iraq. In June Algeria took over the operations of selected oil companies including Shell and Phillips, and in July it raised the taxes on the French companies by 72 cents per barrel without any prior consultation.

The events in Algeria taught Oxy a lesson; by September an agreement was reached. Oxy agreed to pay thirty cents more by raising posted prices from $2.23 to $2.53 per barrel of 40-degree API Libyan crude. The income tax rate was raised from 50 per cent to 55 per cent to 'compensate for underpayment in previous years'. Two weeks later the oasis consortium, whose shareholders included three independents—Continental, Marathon and Amerada-Hess—agreed to the same terms that Oxy accepted. Then Texaco and Socal agreed to the Libyan demands (Penrose, 1975).

It became obvious to the major oil companies that the problem was to prevent the Arabian Gulf producers from trying to outdo the Libyans. Should that happen, the Libyans would make even greater demands and the situation would result in 'leapfrogging' and greater price increases. The companies knew that, in the Middle East, many oil-producing countries' governments were accused of being 'soft' on the oil companies. Thus, for political reasons aside from all economic considerations, the Gulf producers had to ask for at least the gains that were achieved by Libya.

In fact, what the companies feared happened, culminated by a 400-per cent increase of posted oil prices toward the end of 1973. But before we go on, let us see why Libya was successful in realizing its demands.

First of all, the 1967 Arab–Israeli War caused the closure of the Suez Canal and the periodic interruption of the 'Tapline', the pipeline that carries oil from Saudi Arabia to the Mediterranean and that substantially raised freight rates (a very large component of the then prevailing oil price).

Secondly, the Biafra War stopped oil production in Nigeria.

Thirdly, the growing influence of environmentalists in Europe made the low-sulphur Libyan crude more desirable than ever before.

When Colonel Ghaddafi came to power in Libya toward the end of 1969, Europe got 30 per cent of its oil imports from Libya. As a result of these combined factors there was a strong demand for the short-haul Mediterranean oil.

Fourthly, a large quantity of Libyan oil was processed by the small 'independents' who had no alternative sources of supply to honour their previous contracts and to keep their refineries in operation.

The majors refused to supply the independents at posted prices when Libya ordered the cutbacks. That seems to show that posted prices were lower than the then prevailing or expected future market-determined prices. Furthermore, Oxy shares were a 'hot stock', and what made them hot was the Libyan crude. Losing that crude would result in a large capital loss, i.e., the market value of the stock would fall and so would the value of the company's total assets (Sampson, 1975).

Fifthly, Libya was strongly supported by the most militant Arab members of OPEC, Algeria and Iraq (Penrose, 1975).

Many of the Arab governments were convinced that the posted price was less than the real price. Colonel Ghaddafi, a fanatic anti-Communist but also a zealous pro-Arab nationalist, was convinced that the oil companies had no choice but to accept his demand.

James Akins, the then oil expert in the US State Department, reported that 'a top official of a major oil company seriously urged the American Government to dare the Libyans to nationalize'. That official argued that the Libyans could be blacklisted and thus prevented from selling their oil. It was pointed out to the 'top official' that the Libyan Government had $2 billion in currency reserves that would keep it going at current levels of expenditure for four years. And it was also understood by the State Department that the chances of persuading the government-owned European and Japanese oil companies to agree to the blacklisting were rather slim (Akins, 1973).

The Libyan action made it clear to other producers that it was possible to get a higher share of the total oil–revenues. Domestic political pressure on the Gulf producers to demand from the companies (at least) what Libya had won was great.

At the OPEC meeting in Caracas, December 1970, OPEC ordered the oil companies to negotiate with the Gulf producers for higher prices. The companies wanted to negotiate as a united front with all the OPEC members. OPEC refused, and insisted that two agreements should be reached: one between the companies and the Mediterranean producers, and the other between the companies and the Gulf producers. With pressure from the US State Department, the companies finally agreed to start negotiations with the Gulf and Mediterranean producers (Akins, 1973).

M. A. Adelman has criticized the State Department for its 'pro-Arab' stand. Why did the State Department pressure the companies?

The oil specialists in the State Department were concerned about the US dependence on OPEC oil and about the possible political troubles within the producing countries if the companies refused to cooperate. US oil production had peaked in 1970, and an increasing amount of its demand was satisfied through imports from the Middle East. The State Department felt that the US interest would be better served by paying more for oil if doing otherwise would increase the power and influence of the radical forces in the Middle East—as would have happened if the OPEC governments did not get important concessions from the companies.

At any rate, the Gulf producers, after meeting in Teheran on February 3–4, 1971, made a decision to set a deadline, namely February 15, for the companies to comply with minimum terms for higher prices. If there were no compliance, these minimum terms would be enforced by legislation. On

February 12, an agreement was signed. The posted price was raised from $2.18 a barrel, and the tax rate was increased from 50 per cent to 55 per cent (Akins, 1973).

With Libya acting as a negotiator for the Mediterranean countries, an agreement was reached (with great difficulty) on April 2, 1971. The posted price of the Libyan crude was raised from $2.54 to $3.30. Other benefits were also granted by the Tripoli Agreement.

The Teheran and Tripoli agreements were intended to stabilize revenues to the producers and oil supplies to the companies. Thus they provided for 2.5 per cent annual escalation in posted price to offset the inflation of the prices of OPEC's imports from the West. The terms of the two agreements were to be binding for five years.

But by the summer of 1971, the US dollar had sharply depreciated in relation to other currencies, and the US suspended its convertibility into gold. The average inflation rate in Europe and the US was around 9 per cent. The militant members of OPEC felt that inflation was eroding their oil revenues by about 6.5 per cent (9–2.5), and demanded new concessions from the companies.

Thus, in January 1972, an agreement was reached between OPEC and the companies that provided for an 8.59 per cent increase in the posted prices.

However, some OPEC members argued that their income per barrel was low when compared with the excise taxes which Europe levies on its fuel. At the Arab Oil Congress, held during the summer of 1972 in Algiers, 'OPEC was castigated for having been too soft, for having yielded too easily and readily to company and consumer governments' pressure'. The moderate Arab members of OPEC, especially Saudi Arabia and UAE, were on the defensive (Resolutions of the 11th Arab Oil Congress).

In this mood, Saudi Arabia, who could not completely ignore the accusations of the rest of the Arab world even if unjustified, suggested through OPEC the idea of 'participation' to the oil companies. Participation meant the gradual acquisition by the producer governments of defined percentage shares in the producing operations and assets of the international oil companies.

The companies strongly protested, and the US State Department supported them since the Teheran and Tripoli agreement terms, which mentioned no participation, were binding until 1976. But the political atmosphere in the Middle East, due mainly to the Arab–Israeli conflict, made it impossible for the moderate governments of the Gulf to demand any less. The Iraqi and Libyan governments wanted the Arab oil producers to nationalize the entire oil industry, but other Arab countries preferred total ownership of the oil industry through gradual acquisition of the companies' assets. The companies and the State Department were aware of the political difficulties of the producers.

After difficult and long negotiations, an agreement was reached in Riyadh by the end of 1972 providing for the producing countries to acquire percentage shares starting at 25 per cent and working up gradually to 51 per cent. What in fact happened, thereafter, was that all of OPEC's members acquired 100 per cent of their oil operations.

2.5 THE LAST HURRAH: THE QUADRUPLING OF PRICES

Following the 1971 OPEC success in reversing the 1960s trend by raising the posted price of oil and the per-barrel tax paid to the exporting countries, the prevalent view of most of the oil market observers was that an irreversible upward movement of oil prices had started.

Professor Adelman disagreed, 'but the conclusions of this study (*The World Petroleum Market*, 1972: 1) are that crude oil prices will decline because supply will far exceed demand even at lower prices'. What happened to invalidate Adelman's conclusion?

Obviously supply did not increase so as to exceed demand at current prices. The reason for that was the change in the cost of supplying oil that Professor Adelman either ignored or did not anticipate. The meaning of this statement will be completely clear when the reader finishes this book. At this stage it may suffice to say that the companies and the host countries faced different economic constraints which implied different rates of oil output.

Many people in the oil-producing countries were convinced that the companies' rates of output were greater than optimal, even in a country with as much proven and potential reserves as Saudi Arabia. Due to the falling marginal rate of return on domestic investments and the rising inflation rates, increasing oil production became the most unpopular thing for any OPEC member government to do. At this time (1980) a large number of the Arab world's newspapers are calling for drastic cuts in production for conservation purposes.

The environmentalist pressure in the consuming countries also had led to an increase in the rate of substituting oil for other sources of energy.

For whatever reason, planned demand for oil rose faster than *actual* supply in the period between 1971 and 1973 (*Petroleum Intelligence Weekly*, August 20, 1973). As will be explained in greater detail in another chapter, the realized market price rose faster than the posted price—a reference price that was determined by negotiations between OPEC and the companies.

As early as May 1972, Saudi Arabia sold all of its participation crude available for 1973 at 'record prices'. The exact figure was not made public, but the *Middle East Economic Survey* (May 11, 1973) reported that it was 'much higher' than the posted price. In June, Qatar sold all of its participation crude for the period of 1973–1975 at a price 'in excess' of posted price. By August 20, *Petroleum Intelligence Weekly* concluded that realized prices in the Mediterranean had gone to 'well above' posted prices.

In June of 1973, it was calculated by one of OPEC's economic experts that, since 1971, prices of crudes offered for sale in arms-length deals had increased by 65 per cent compared to an increase of 33 per cent in posted prices on which governments' take was based (Kubbah, 1974). (Arms-length deals refer to transactions that involve a third party rather than the usual exchange between the producing company and its host government.)

On the 6th of October, 1973, the third Arab–Israeli War started. During that war, almost every Arabic newspaper that reflected public opinion demanded the nationalization of the American oil companies and an embargo on oil exports to the US as a penalty for US 'war-like behaviour' towards the Arabs. The Arab governments had to do something. Accordingly, the oil ministers of the six Gulf countries (including Iran, a non-Arab nation) met in Kuwait on October 16 and decided to fix new crude oil prices unilaterally. Still short of an explicit nationalization but for *the first time* in the history of OPEC dealing with the oil companies, *legislation had been substituted for negotiation*. The posted price was raised from $3.011 to $5.119 per barrel, an almost 70 per cent increase.

On October 19–20, the members of OAPEC met in Riyadh and declared an embargo on oil exports to the US and the Netherlands. It soon became realized that an embargo alone, against anyone, was not effective since there were non-Arab producers and the oil companies could simply shift customers. For an embargo to have any effect, a shortage of supplies at the going prices had to be induced so that oil consumers would compete for the limited oil quantities.

Sure enough, OAPEC learned the underlying economic principles fast and, on November 5, ordered the oil-producing companies in the Arab world to cut production by 25 per cent of the September level and 5 per cent each successive month until the US adopted 'a more sympathetic attitude' towards the Arabs.

The market reaction, as predicted, given the price inelasticity of crude oil supply and demand in the very short run, was very favorable to oil sellers. This means that the percentage change in quantities of oil supplied or demanded is much smaller than the associated percentage change in the price. The reason for the slow change in quantity of oil demanded is that it takes time for the users of oil to either find a substitute for it or adjust their factories, planes, automobiles, etc., to use less of it. Similarly on the supply side, it takes time to find new oil or expand the output of already known oilfields.

Reflecting the independent oil companies' scrambling to obtain the crude and the majors' attempts to build inventories in anticipation of smaller future supplies, the free market oil price rose very sharply.

When the Iranian Government called for bids, on December 10, for the 0.5 million b/d that it intended to market in the first six months of 1974, it received bids that reached from $16 to over $17 a barrel (Kubbah, 1975).

That was the first government sale of crude since OAPEC cut back output. The Iranian government wanted to test the market and, instead of the hitherto followed practice of simply announcing the price at which it would sell, it asked for bids in an attempt to determine by how much OPEC could raise its posted price.

The Iranians had long argued within OPEC that the posted price for oil should equal the cost of the nearest substitute. The OAPEC restrictions on oil output strengthened the OPEC members, such as Iran, who wanted dramatic price increases. It is worth emphasizing that the embargo and the reductions of output were determined within OAPEC and not within OPEC. As will be shown in the other chapters, the posted price increases that were subsequently made by OPEC would have been most likely possible without any OAPEC embargo.

At the Teheran OPEC Ministers' meeting on December 22–23, 1973, the Iranians demanded that the posted price should be raised to $16, the price that was received for their direct sale of oil. To make the Iranian argument stronger, the Nigerians and the Libyans reported that they received bids of over $20 a barrel for their closer-to-Europe, better-quality oil.

The Saudi Arabian oil minister resisted the huge price increase on the grounds that the high bids did not reasonably represent appropriate market prices due to the OAPEC output cutbacks and destination restrictions. The Algerian, Iraqi, and Libyan ministers took the Iranians' position and added that if output reductions increased prices and improved the terms of trade in favor of the oil exporters, so be it. A compromise was reached, and the posted price was raised to $11.65 a barrel for the Arabian light 34° API.

Thus, the posted price prevailing before the outbreak of the 1973 war nearly quadrupled. But more importantly, the era of determining the oil price by negotiation between the companies and the host countries was gone for ever. The most important event, *that had drastically changed the international oil market, took place on October 16, 1973, when the oil producers resolved to substitute legislation for negotiation. The net effect was a shift of property rights in crude oil from the private western oil companies to the host countries.*

Almost all the writers who claim that OPEC is a cartel (an agency that monopolized the oil market) cite the movement of posted prices—which until October 1973 reflected the bargaining power of the companies *vis-à-vis* the host countries rather than market forces—as evidence without appreciating the influence of the change of property rights in crude oil on the international oil market. Whether OPEC is in fact a cartel or not will be the subject of the following chapters.

REFERENCES

Adelman, M. A. (1972). *The World Petroleum Market*, Johns Hopkins University Press, Baltimore.

Akins, J. A. (1973). 'The Oil Crisis: This Time The Wolf is Here', *Foreign Affairs*, 463–90.
Arab League (in Arabic), (1973). *The Efforts of the Arabs in Oil Affairs*, Cairo.
Chandler, G. (1973). 'Some Current Thoughts on the Oil Industry', *Petroleum Review*, 6–12.
Jacoby, N. H. (1974). *Multinational Oil*, Macmillan Company, New York.
Kubbah, A. (1975). *OPEC: Past and Present*, Petro–Economic Research Center, Vienna.
Middle East Economic Survey (MEES), (May 11, 1973).
Mikdashi, Z. M. (1972). *The Community of Oil Exporting Countries*, Cornell University Press, Ithaca, New York.
OPEC, (1976). *The Structure of the Organization of the Petroleum Exporting Countries*, OPEC, Vienna.
Penrose, E. T. (1975). 'The Development of Crisis', *Daedalus*, 39–59.
Penrose, E. T. (1969). In *The Petroleum Industry*, Hearings Before the Subcommittee on Antitrust and Monopoly of the Committee on the Judiciary, U.S. Senate, 91st Congress, 1st Session, Part I, 432–33.
Petroleum Intelligence Weekly, (August 20, 1973), 35.
Resolutions of the 11th Arab Oil Congress (in Arabic), (1973). Algiers.
Rouhani, F. (1971). *A History of OPEC*, Praeger, New York.
Sampson, A. (1973). *The Seven Sisters*, The Viking Press, New York.

CHAPTER 3
The Stability of Cartels and Price Leadership

3.1 PRELIMINARY REMARKS

The nature of demand and supply—and occasionally government policy—make the market for a commodity to be characterized by a high degree of monopoly or high degree of competition. Perfect competition and pure monopoly rarely exist in the real world, if the two concepts are properly understood, but they are frequently used as fables for economic analysis to reduce the complexities of the real world to a manageable level.

Further, these two concepts, though in no way accurate descriptions of reality, have proved very powerful in predicting the behavior of market participants. In fact the market for the largest number of commodities is characterized by some degree of competition and some degree of monopoly. What we want to explain now is why in a market for some goods a greater degree of competition prevails and for some other goods a greater degree of monopoly predominates.

Competition will dominate, in the absence of government interference, a market for a good if the size of the market and the technology of its production will allow the coexistence of many producers, each of whom could not increase his wealth by producing something else.

The exact number that will assure competition can not be determined *a priori*, because it depends on the size of the market and the technology of production. Five grocery stores may assure competition in a small local neighbourhood and may lead to monopolistic practices in a big metropolitan area.

The essence of competition, as George Stigler has observed, is not rivalry but rather the utter inability of any individual seller to influence the market. Competition is expected to 'discipline' the market participants to produce in the cheapest way and sell at the lowest possible prices. Anyone who does otherwise will either forego some extra wealth he could have realized or simply go out of business (Stigler, 1968: Chapter 5).

Monopoly will prevail in a market when either the size of the market will allow only one producer to exist or when the technology of production is such that the bigger the firm (or plant) the smaller the cost of production per unit of output. (Of course cost per unit will never fall for ever, but may continuously fall up to that level of output which satisfies the entire demand in a given market.)

Monopoly will lead to higher prices and sometimes also to higher costs per unit of output. The expected behavior of a monopolist is a consequence of the absence of the discipline of competition. It is important to note that monopoly does not assure large profits or necessarily imply bigness of the monopolist firm. Profits depend also on demand conditions; the single butcher in a small village has a monopoly power over the retail prices of meat in his village.

There are, however, a significant number of markets where the size of the market and/or the technology of production will only permit the coexistence of few producers (or sellers). In such markets some of the individual sellers can influence the market price by increasing or decreasing sales. That does not mean that the big seller does as he pleases. It means he may increase the price to a certain level without losing all of his sales or he may reduce his price to a certain level without causing all or some of his rivals to sell nothing.

It has been claimed that one seller may be so large that he can charge a higher price because each of the rest of the sellers is so small and takes the big seller's price as given. If that is the case then either the price the big seller is charging is not above its competitive level (otherwise the small sellers will expand their outputs) or he has cost advantages which could not last permanently unless he is getting a government subsidy, or the other sellers could not expand their outputs because of government restriction. When cost advantages are due to access to a better quality natural resource then the price of the product reflects the rent payments for the use of the superior resource.

The fundamental difference between markets where no individual producer could influence prices and those where he can is the speed of adjustment of prices and outputs to changes in demand and/or costs of production. In the case of the large number of sellers, each may be a price taker, and prices and output automatically get adjusted when the forces underlying demand or costs of production change. In markets of few sellers, normally no two firms produce exactly the same product—a Ford is not *exactly* a Chevrolet—and as a result each seller has some monopoly power over its product. That is, a producer could raise his price by a small amount and still sell some and reduce his price by a small amount without capturing all the market. Of course if his price changes are large enough he may sell nothing or all he produces. Each of the sellers is a price searcher; when demand falls all he knows is that his sales fell. He waits to find out if it is more profitable to reduce prices, or reduce outputs, or increase inventories.

To recapitulate, whether finally competition or monopoly will dominate a given market depends on economic forces rather than on the desirability of one form of market structure over another. It is intuitively expected and formally has been documented that monopoly is preferred by sellers over competition.

That is to say sellers, whether one or ten million, would like to curtail competition if they can. The extent to which competition prevails is due to the sellers' inability to eliminate it. In many cases cooperation among sellers to reduce competition is unthinkable; for example, their number may be so large that the cost of merely contacting all of them is many times greater than the expected increase in wealth that cooperation will make possible.

Nevertheless, it is very reasonable to expect that when the number of sellers is so small that each of them is aware of the existence and economic actions of his rivals, attempts will be made to reduce competition.

The potential for cooperation among sellers does not, however, assure that it will occur. In fact it almost never does (at least for a long period of time). Careful analysis has shown that the famed European cartels, particularly the German ones, existed only because government action made it possible for them to exist (Stigler, 1968). It is to this question of instability that we now turn.

WARNING: The above paragraphs may suggest that market adjustments—to changes in costs and demand—occur instantly. They do not. In the process of adjustment some sellers increase their wealth unexpectedly, and some lose in the same manner. But the ability of a firm to predict or at least correctly interpret changes in demand and costs is what finally determines its survival in the long term. An element of chance may exist, and thus some firms, for a short period of time, may get 'ahead' of others through no special talent or greater effort on the part of their management.

3.2 THE DIFFICULT TASK OF COLLUDING

Sellers' attempts to reduce competition may take many forms, but the basic problem, that always makes life difficult for all kinds of colluding groups, is still the desire of each of the sellers to increase his gains at the expense of, or without regard to the interest of, the rest of the group.

Our main concern here is collusion that takes the form of a cartel. For the purpose at hand, a *cartel* refers to an agency that makes pricing and output decisions for its members in order to monopolize the market. (An ideal cartel would establish monopoly price and rate of product output that maximizes the wealth of the entire group of sellers or producers.)

Agreements on a monopoly price are relatively easy, though not trivial or costless by any means. The problem lies in finding a *satisfactory* formula that ensures the permanent maintenance of such a price. One possible formula that a cartel may adopt is to set production quotas for each producer so that the total output of the group will assure the monopoly price for the product.

But because this market price is a monopoly price, each seller will attempt to increase his output, because such action will add more to his revenues than to his costs. An individual seller will take the market price as given and thinks that he could sell more than his quota without reducing the market price. Unfortunately for him, every other member of the cartel faces the same temptation to increase his output. The net result is that supply will rise and consequently the price will fall.

The cartel may try to detect cheating and punish cheaters. Policing collusion is costly, however, and the greater the rewards for cheating, the higher the costs of policing (Stigler, 1968). Conceptually it is possible for cartels to be formed and thrive only as long as the cost of detecting and punishing the 'fifth column' of cheaters is less than the gain of collusion.

Empirically, however, profitable cartels did not last, in spite of government help, precisely because the gains from cheating as viewed by individual producers were too great. Unprofitable cartels, like the IATA, which attempts to regulate the world air carriers, are able to last (though the IATA is now in the process of collapsing since government regulations have allowed entry by innovating entrepreneurs like Laker) precisely because the rewards for cheating were not great enough.

In short, the greater the amount by which a cartel price will depart from the competitive price the greater the incentive for its members, acting individually, to cheat by increasing their outputs and thereby destroying the cartel price.

One may conclude, as Alchian and Allen did, that '... the attempt to remove market competition does not eliminate competition. It shifts its form or locale—in the case of collusion, to the conference table, where the competitive issue is the division of the gain—by no means an easy one to resolve' (Alchian and Allen, 1974: 354).

But, one may wonder, does not the emergence and the apparently lasting strength of OPEC fly in the face of our analysis? No, it does not, claims to the contrary notwithstanding. I will argue that, whatever OPEC may be, it is *not* a cartel.

3.3 COLLUSION WITHIN OPEC

Since October 1973, the 'OPEC Cartel' has claimed the headlines mainly because of the importance of oil in the world economy. But OPEC also attracted the attention of some seasoned economists—and some others who are not so seasoned—precisely because it *appeared* that OPEC had emerged as a very effective and successful cartel.

OPEC is thought of as a cartel mainly because of the suddenness and magnitude of the oil price changes that occurred between January 1973 and January 1974.

In this book I will advance an alternative hypothesis, which would account for the exceptional success of the 'OPEC Cartel': that OPEC is not a cartel at all, but that other structural changes explain the changes in the price of oil in the world market. But that is getting ahead of our story. At this point we will confine our analysis to the evidence that OPEC is not a cartel rather than to any alternative hypothesis for explaining the oil market behavior.

Why do we say that OPEC is not a cartel? To begin with, OPEC does not set any output quotas for its members. No individual member is compelled to abide by OPEC's pricing decision. No side payment in any form is made. If the gains that OPEC is supposedly making are made possible by eliminating market competition, then, how are those gains being distributed? Either the OPEC price is not a monopoly price or someone is doing something to prevent acts that normally erode cartel prices.

Some commentators have suggested that what keeps OPEC together is Saudi Arabia by its individual output cutbacks. As I will argue in the next chapter, the actions of Saudi Arabia seem to indicate that its output behavior is designed to preclude the oil price from rising rather than averting its fall.

3.4 WHY CARTELS?

Economic theory teaches us that when we are able to determine that a price of a commodity persists well above its marginal cost, then its producers must have some monopoly power. The source of monopoly power must be either decreasing costs (natural monopoly) or agreement among producers to act together in collusion. It should be noted that barriers to entry could explain why the number of producers does not increase, and they could explain why average costs could be lower than what they are, but they could not explain why the existing firms do not expand output to a point where price (P) equals marginal cost (MC). And if the marginal product (MP) increases as the rate of output rises, then we have a continuously decreasing MC, which is the natural monopoly case, and not any other barrier to entry. In effect, when monopolistic practices are present in the absence of continuously decreasing costs, then collusion must be present.

The economic incentive for firms to collude is shown in Figure 3.1. Suppose we have an industry composed of N firms. The demand the industry faces is $D(P)$. The industry's supply is $S(P)$ and is equal to ΣMC_i. The competitive solution is given by the intersection of the industry's $S(P)$

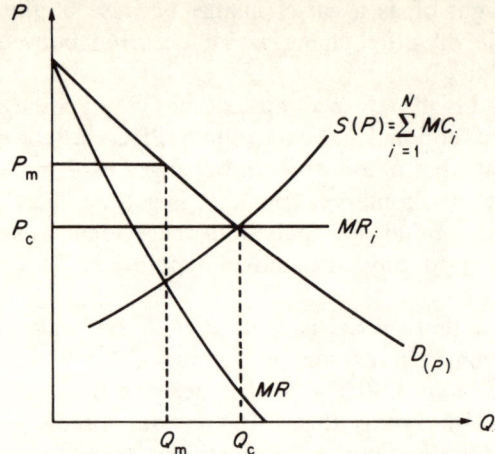

Figure 3.1. The industry's incentives to collude.

and $D(P)$ and results in the equilibrium price (P_c) and the equilibrium rate of output Q_c.

Although this competitive solution was achieved as a result of each individual firm's efforts to maximize its profit, it does not lead to the maximization of the industry's profits (π) because, for any individual firm, its MR_i is equal to P_c. For the industry, $P_c > MR$ at every rate of output. As each firm produces more to equate its MC_i to its MR_i (which it takes as given), it reduces the revenues of other firms.

If firms could agree to act together, then the rates of output and price that maximize the total industry's profits are Q_m and P_m where $MR = MC$,

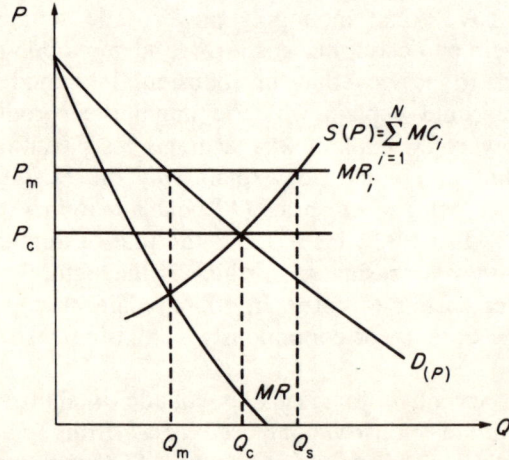

Figure 3.2. The incentive for individual firms to depart from the collusive solution.

the monopolistic solution. The industry's profit is maximized if its N firms agree to reduce output to Q_m instead of maintaining Q_c.

In practice that is not an easy task. Just as it is more profitable for all firms together to depart from the competitive solution, it is equally more profitable for each individual firm to depart from the monopolistic solution. Figure 3.2 shows why.

At the monopoly price, P_m, an individual firm sees its MR_i to be equal to P_m rather than MR, and thus would maximize its individual profit π_i by producing part of Q_s. Every firm has an incentive to increase its rate of output unless there is an effective mechanism of allocating outputs and profits. The collusion must be policed, and that is not free (Becker, 1971: 98–101).

3.5 A THEORY OF CARTELS

Let us say that collusion increases revenues by R_m and its costs to form and police by C_m; then collusion will take place if $R_m - C_m > 0$. What determines R_m and C_m?

The magnitude of R_m is determined by the difference between marginal revenue and price. That is,

$$R_m = P - MR$$

But

$$MR = P\left(1 - \frac{1}{e}\right)$$

where e is the price elasticity of demand. Then

$$R_m = P - \left[P\left(1 - \frac{1}{e}\right)\right]$$

$$R_m = \frac{P}{e}$$

or

$$\frac{\partial R_m}{\partial e} < 0$$

This says the industry's gains from collusion are negatively related to the elasticity of its demand curve.

When it comes to the costs of forming and policing a collusion, the situation becomes very complex, yet we can make some generalizations (Stigler, 1968: Chapter 5).

First, the number of firms in the industry. The larger the number of sellers, the more difficult it becomes for them to agree on the price structure and the greater the probability that cheating will not be detected.

It becomes difficult to know who is doing what. On the other hand, the smaller the number, the greater the influence of the action of any individual member on the rest of the group. Recognizing interdependence reduces the probability of chiselling since each potential cheater will know that its rivals will detect its violations and would either punish it or do the same, resulting in no net gains for a chiseller.

Second, homogeneity of the product. The more complex the industry's product structure, the greater the chances of disagreement among sellers on price structures. But when the elasticity of substitution between any pairs of firms' products is infinite, product differentiation becomes impossible, and the only way for an individual producer to increase its net return is to collude with others. For example, American wheat producers would gain from collusion if the costs of collusion for them (mainly because of their large number) were not prohibitive.

Third, number of product buyers. The larger the number of buyers, the more difficult it is for any one seller to increase his share of the market (through advertising, price reductions or product improvement, credit terms, etc.) without the knowledge of his rivals; thus Stigler concludes: 'Chiselling will be harder to detect the fewer and larger the buyers.' Of course, if the buyers are government agencies and official records are made public, then detecting chiselling will become almost costless.

Fourth, customer loyalty. The less erratic the shift of customers among sellers, the easier it is to detect chiselling and the smaller the chances of competitive price cutting.

We can formalize the foregoing discussion of the costs of collusion in mathematical language. Let

N = the number of firms in the industry
H = product homogeneity
B = number of buyers
L = customer loyalty.

Then,

$$C_m = C_m(N, H, B, L)$$

If we define the net gain from collusion as

$$G = R_m - C_m$$

then it follows that

$$G = G(e, N, H, B, L)$$

where

$$G^1 < 0, G^2 < 0, G^3 > 0, G^4 > 0, G^5 > 0$$

It should be noted that we are tactically assuming that $dC_m/dR_m = 0$. That is, cross derivatives are zero. In other words, we assumed, for simplicity, that the factors that determine R_m are independent of the factors that determine C_m.

The preliminary conditions are

1. If $G = R_m - C_m \geq 0$, then collusion will occur
2. If $G = R_m - C_m < 0$, then collusion will not occur

But, in the long run, members of collusive agreements must worry about production by non-members. The price they charge for their product should not be so high as to raise their gains from collusion by an amount that encourages production by non-members. That is, in the short run some increase in gains from collusion will be foregone for long run optimization.

Therefore, R_m should have an upper limit, say θ. For collusion to occur and continue to exist, two conditions must be satisfied:

1. $G = R_m - C_m \geq 0$
2. $R_m \leq \theta$

In other words, what this model is saying is that net returns from collusion are negatively related to the elasticity of the industry demand curve and positively related to the elasticity of the industry marginal cost curve. If the costs of collusion are large relative to the gains from collusion it does not pay to collude; and, conversely, if costs are low relative to gains, then collusion is profitable.

The significant point that emerges from these analyses is that the competitive or the monopolistic solutions could be viewed as outcomes of costs and benefits of collusion. Namely, if the benefits of collusion are greater than the costs then collusion will occur. If, on the other hand, the costs of collusion are greater than the benefits then competition will prevail.

Every collusive agreement, whether in perfect form (e.g. centralized cartel) or an implicit understanding, creates at least two problems for its members: (1) an external problem—how to predict and, if possible, discourage production by non-members; and (2) an internal problem—how to share the net gains from collusion and how to prevent these gains from falling to zero due to the costs of policing.

3.6 OPEC AND COLLUSION

Because of the absence of a central market for oil there is a difficulty in knowing what the exact price of a unit of oil is at any time. The posted prices of oil have always been made public in the fifties and sixties by the oil companies and in the seventies by OPEC, and as a consequence figures on posted oil prices are easily available, but it is not always certain whether these prices reflected real market forces. This ambiguity should be kept in mind when we discuss the price of oil.

Before the 1971 Tripoli and Teheran agreements the total cost of oil was believed to be around £5 a ton delivered to Western European markets. At that time a ton of coal cost an average £7 to £8 to produce in Britain. Since it takes 1.5 tons of coal to match the energy equivalent to one ton of oil, the cost of oil's nearest substitute was £10 to £12 a ton. The cost of a ton of coal from marginal pits was considerably greater (*The Economist*, April 26, 1975: 36).

It was obvious that a substantial oil rent existed. The oil rent could be obtained by pricing oil up to just below the price of the cheapest substitute such as coal. This rent could have been realized by producer countries as a monopoly profit (a modest fraction was).

As a result of the producers' failures to 'cream off' a greater part of the oil rent, the consuming countries' governments got a greater part of this rent in the form of taxes from oil consumers. Some of the rent was taken by the international oil companies, who produced more expensive oil outside the OPEC countries.

Before the so-called oil 'crisis' of December 1973, the price per barrel of refined oil products in Western Europe was about $14.50. The consumer government's take of the $14.50, in the form of taxes, was 51 per cent ($7.395) while the share of the producer governments was only 16 per cent ($2.32); the rest of the price of a barrel of oil product went to the oil industry in the form of costs and profits ($4.785 or 33 per cent) (*The Economist*, April 26, 1975: 37).

These figures show that the rewards for collusion (if it could be effectuated) on the part of the oil producers could be substantial. What did really happen? Did collusion occur? In short, is OPEC a cartel?

It may appear that I am belaboring the obvious, since everyone—almost everyone—says OPEC is a cartel. Dr. Paul H. Frankel a well-known specialist in the petroleum industry, wrote, 'OPEC is what American authors still call by way of oversimplification "a cartel".' (Frankel, 1973). Professor (of Economics) Zuhayer Mikdashi of the American University at Beirut goes further. 'Officials and scholars of several developed market economies,' he wrote, 'have chided developing countries' attempts at concerted action—often branding them as cartelists or exploiters. Such an accusation is unfounded, even for the successful grouping of oil-exporting countries' (Mikdashi, 1974). But neither Frankel nor Mikdashi gave any reason why OPEC was not a cartel.

3.7 OPEC AS A JOINT SALES AGENCY

When each member of a collusive agreement completely surrenders its power to make management decisions and functions to a joint sales agency, we have a case of perfect collusion or what is sometimes referred to as a 'centralized cartel'.

In the case of a centralized cartel, the decision-making with regard to pricing, output, sales, and distribution of profits is accomplished by the central agency. The centralized cartel can maximize the entire industry's profit by producing the rate of output at which the industry's *MC* equals the industry's *MR*. And maximization of the industry's profit becomes, essentially, a multi-plant monopolist problem (Stocking and Watkins, 1948). Each firm should produce the rate of output that equates MC_i to the industry's *MR*. (This result is too familiar to be derived here.)

Collusion models are simplified abstractions from more complex actual cases that are not designed to describe reality but to predict, with varying degrees of accuracy, the outcomes of collusive agreements.

The economists who in 1973 predicted that OPEC would immediately crumble were obviously thinking about the difficulties of reaching agreements on profit sharing and production quotas, for they knew that the individual oil producer would not surrender its economic decisions to the OPEC organization.

In fact they were right: Cartels do crumble; they were wrong, however, in thinking that OPEC was or is a cartel. To start with, all attempts (within OPEC) toward production quotas and demand pro-rationing were abandoned in the early 1960s, long before OPEC had been thought of as an effective cartel that eliminated competition in the world oil market. A satisfactory formula for output quotas could not be found. Venezuela, then the largest producer, wanted historical levels of output to be used as the base, Iran favored population, and Saudi Arabia and Kuwait selected proven reserves as the base for output quotas (Mikdashi, 1972: 30-68).

At this time OPEC does not set any output quotas for its members. (Obviously, output is being restricted—when compared with its historical levels—to keep the price at its current level, but we will come to that later.) When OPEC finally acquired the power to set prices unilaterally (without any negotiating with the oil companies) in 1973, it did not impose compulsory requirements on its individual members to abide with its pricing decisions. The agreements on prices are strictly voluntary. No rewards are bestowed on abiders or penalties imposed on the violators. No side payments of any kind are made (Mikdashi, 1975). Some Arab members of OPEC make regular transfer payments to Syria, Egypt, and Jordan, but none of the receiving countries is a member of OPEC.

The political and economic differences among OPEC members made it very difficult, if not impossible, for the OPEC organization to act as a central cartel with power to set prices and allocate production quotas. The function of OPEC, as an organization, is limited to collecting technical and economic data. The pricing decisions are made by the oil ministers who represent the different countries. But setting prices is not enough to assure that such agreed-upon prices will prevail.

The producers of many other raw materials had met many times to set prices. The problem is that they rarely abide by the agreed-upon prices. Thus in the case of world oil, either the price is not a monopoly price, or something else—very mysterious and unknown to other hitherto known colluding groups—is making OPEC such an exceptionally successful device for eliminating market competition!

3.8 PRICE LEADERSHIP BY A DOMINANT FIRM

It has been supposed that when an industry is characterized by the coexistence of one (or few) large firm (or firms) and a large number of small firms, the large firm(s) will set the market price. It is alleged that the large producer (or seller) is aware of its economic power while the smaller ones acting individually could not influence the entire industry's product price or rate of output.

Under such circumstances it was thought that the large or dominant firm will consider its demand to be the difference between the entire market demand and the total collective output of the other small firms. For the dominant firm to maximize its profits, it will produce the rate of output that equates its marginal revenues (given its demand) to its marginal cost.

That means in plain English that the large producer will choose the rate of output at which the change in his revenues is equal to the change in his costs. If the addition to revenues of producing one more unit is greater than the addition to costs, then output should be expanded; and conversely if the addition to revenues is less than the addition to costs.

Each of the small firms is supposed to consider the dominant firm's chosen price to be the market price. Because of its smallness it could not possibly hope to change that price. Thus it tries to maximize its wealth given its own costs and the dominant firm's price.

Many readers, however, will not find the above analysis to be flawless—not by any means. To start with, one may wonder how did the dominant firm become so big while the other firms remained so small? If that occurred because the large firm was in the market first then the large firm's dominance is temporary, and it could not exercise much monopoly power before the others caught up with it. That is to say, if the nature of the technology of the product favors larger plants then the small firms will also expand.

One may think of the limited access to a natural resource that will give a firm an economic power that is being denied other firms. Although that is conceivable one would be hard pressed to give an example of a single natural resource which is accessible to a single private producer and for which there is no close substitute. The more practical explanation for the emergence of a dominant firm is government policy that favors one firm at the expense of other actual and potential competitors.

One of the most important things that may favor a large producer is the cost of collecting and analysing market information. In a large number of markets the costs of knowing who the customers are, who they will be in the future, and forecasting future technological changes that affect the nature of production and a host of other things, may constitute a significant part of the total cost of production. And there is no doubt that the larger the rate of output the smaller the cost per unit of market information. But before we jump to conclusions let us remember that the advantages that bigness allows when it comes to the cost of information do not prevent the emergence of costs that result from bigness. Among such costs are the cost of coordination and the flow of necessary information within the firm. It is rather obvious that the costs of coordination and dissipation of information within the firm are smaller the smaller the firm.

If a dominant firm exercises price-leadership as a tactic for collusion, then we still have the problem of dividing the gains from collusion. A problem which, as was mentioned before, is not easy to contend with.

We may summarize this section in the following statement: Models of price-leadership by a dominant firm are strictly short run analysis. That is, they may describe reality in cases when not enough was known and/or when more time is needed for adjustment by the existing small firms, or entering of new ones. Given enough information and/or time, the smaller firms will expand, get out of the market, or simply coexist with the large producer when neither bigness nor smallness will result in any clear advantages. Barring government help to a given producer, the theoretical basis of price-leadership by a dominant firm simply is not as strong as has been frequently supposed.

Yet, one may wonder, 'does not Saudi Arabia dominate the world oil market?' As we will try to show in other parts of this book, Saudi Arabia did have a great influence on the world oil market. But it did so because oil is a non-reproducible resource and because Saudi actions were designed to keep the price lower than what it otherwise would be.

In the short term Saudi Arabia was capable of exercising price-leadership precisely because it was trying to keep the price lower than other oil producers wanted it to be, rather than higher. The task of price-leadership would have been more difficult, if not impossible, had Saudi Arabia wanted higher prices than other 'smaller' producers wanted. In the latter case there is not much that Saudi Arabia could have done to prevent small producers from expanding their output other than shouldering the brunt of the output cutbacks, to allow other producers to expand without a glut and price reductions. Saudi Arabia could not have pursued such a course for ever, even if it had wanted to.

The Iranian 'crises' of the autumn of 1978 and the winter of 1979 showed that what Saudi Arabia could do to prevent sharp rises in oil prices is limited, even though Saudi Arabia did increase its oil output by 1.5 million b/d to avert huge increases in the price of oil.

3.9 THE FORMAL MODEL

Consider an industry with N firms. The industry produces a homogeneous product, Q. The N firms are made up of one large firm and

($N - 1$) small firms. For simplicity's sake, we assume that each of the ($N - 1$) firms is so small that it ignores the effects of its actions on the price of the industry's output and on the decisions taken by the dominant (large) firm or by other smaller firms in the industry.

In other words, smallness in this context means that a firm is small if it faces a perfectly elastic demand curve; and largeness (or dominance) means a single firm is dominant if it faces a demand curve that is not perfectly elastic. Empirically, it is not always obvious whether a demand curve is clearly perfectly elastic or not, but at this stage we will be satisfied with conceptual clarity and shall ignore the empirical ambiguity until we apply the model.

Let $D(p)$ denote the industry demand curve and let $S(p)$ the supply curve of output in the aggregate for the small firms in the industry. Because each of the small firms acts like a perfect competitor, $S(p)$ is the sum of the marginal cost curves of the small firms; that is,

$$S(p) = \sum_{i=1}^{N-1} MC_i.$$

The dominant firm sets price so as to maximize its profits, taking output of the small firms as given by $S(p)$. The demand curve for the dominant firm, $d(p)$, is obtained by horizontally subtracting $S(p)$ from $D(p)$ at each value of P; that is, $d(p) = D(p) - S(p)$. Figures 3.3 and 3.4 indicate the profit maximizing choice by the dominant firm.

The demand curve, $d(p)$, for the dominant firm intersects the price axis at p^*; at any $p \geq p^*$ the small firms will supply all of the output demanded and $d(p) = 0$. For $p < p^*$, the dominant firm demand is the difference between $d(p)$ and $S(p)$ in Figure 3.3. Thus $d(p)$ is more elastic than $D(p)$.

The dominant firm maximizes its profits by equating its MR_d and MC_d at a rate of output Q_d and a price level P_d. When the dominant firm sets a

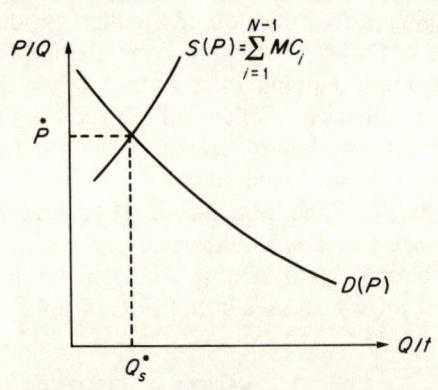

Figure 3.3. Industry demand $D(P)$ and small firms' supply $S(P)$.

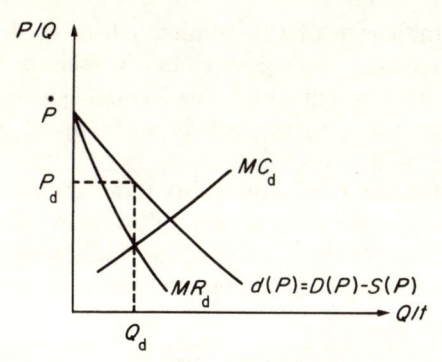

Figure 3.4. Choice of price P_d and output Q_d for the dominant firm.

price level equal to P_d, the $(N - 1)$ small firms supply Q_s, as shown in Figure 3.5. The industry rate of output is $\bar{Q} = Q_s + Q_d$.

The problem of supply indeterminacy in the case of pure monopoly is also present in price-leadership models. The supply curve of our dominant firm is not defined. The major difference, however, between the dominant firm model and the pure monopoly model is that the dominant firm is affected not only by the forces that determine the industry demand curve and the ones that determine its own MC, but also by the forces that determine the marginal cost curves of the smaller firms in the industry. Any change in the slope or the position of the $S(p)$ curve affects the $d(p)$, which in turn changes the profit-maximizing price and output for the dominant firm.

The dominant firm model does not say that price-leadership will occur in every industry which is made up of a single large firm and a cluster of

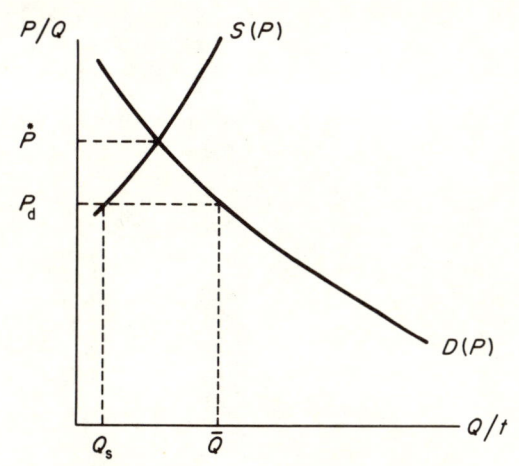

Figure 3.5. Industry equilibrium with a dominant firm producing $Q_d = \bar{Q} - Q_s$.

small ones. The significance of the dominant firm model is that the large firm is aware of its economic power and would exploit it, if it were profitable to do so, through price leadership, given an industry that produces a homogeneous product and is made up of a large one and a collection of smaller ones.

Under such conditions, and in the short term, a single large producer is aware of the effects of his action on his industry product price and rate of output while the other smaller firms, acting individually, could not affect the industry product price or rate of output.

REFERENCES

Alchian, A., and Allen, W. (1974). *University Economics*, Prentice Hall International, London.
Becker, Gary S. (1971). *Economic Theory*, Knopf, New York.
The Economist (April 26, 1975).
Frankel, Paul H. (September 1973). 'The Oil Industry and Professor Adelman', *Petroleum Review*, 347–349.
Mikdashi, Z. M. (1972). *The Community of Oil Exporting Countries*, Cornell University Press, New York.
Mikdashi, Z. M. (Winter 1974). 'Collusion Could Work', *Foreign Policy*, 57–68.
Mikdashi, Z. M. (Fall 1975). 'The OPEC Process', *Daedalus*, 304–328.
Stigler, George J. (1968). *The Organization of Industry*, Irwin, Homewood, Illinois.
Stocking, G. W., and Watkins, M. W. (1948). *Cartels or Competition*, The Twentieth Century Fund, New York.

CHAPTER 4
The Economics of Exhaustible Resources

4.1 THE FORMAL MODEL

Natural resources, whether exhaustible or not, are assets of the society in that they can yield streams of services to the society over future periods. But certain features of natural resources may distinguish them from other assets. These features include the externalities that arise in the production and consumption of services derived from natural resources, the appropriability properties of natural resources, and whether such resources are replenishable or not.

In the case of Saudi oil there is no problem with either appropriability (ruling out foreign intervention) or externalities. The government is the sole owner of all of the country's natural resources. Thus, property rights are well defined, enforced, and transferable, and externalities that arise due to the divergence between social and private costs are absent.

Oil, of course, is exhaustible in the sense that society's stock of the reserve could not be physically replenishable. The difficulty in the concept of exhaustion is that geologists estimate the 'stock' of oil in terms of 'recoverable' reserves. But what is recoverable depends on the cost of extraction relative to the price of oil. Every increase in the price of oil or decrease in the costs of its extraction increases the world's stock of recoverable oil. In addition, an increase in price will lead to more search, and that usually results in increasing reserves.

In a physical sense, all minerals are limited because the crust of the earth is. Shale oil, tar sands, coal, and uranium are all limited. The difference between them and oil is that oil is at this time cheaper to extract and use. When it becomes more expensive than all the others, we stop searching for it and use the others before we get to the last drop of oil that the earth contains. Thus oil is really inexhaustible in a physical sense, but, for the purposes of economic analysis, exhaustion occurs once it is cheaper to use other sources of energy.

For our purposes we will define the exhaustion of oil in the following way (Quirk, 1976: Chapter 18):

Given prices of oil and substitutes and a state of technology in period t, the oil reserves R at $t + 1$ are equal to reserves at the beginning of period t minus actual output q in the tth period, or

$$R_{t+1} = R_t - q_t$$

The problem that an owner of an exhaustible resource faces is how much output should be produced in each period of time until the costs of production exceed the revenues, i.e., exhaustion. To answer this question we have to make some simplifying assumptions.

The mathematics could be challenging and might involve multivariate control problems which are very difficult to solve. Happily, James Quirk has shown that all the important results could be derived by using simple algebra.

Assume that (1) there are many producers and each acts as a perfect competitor, (2) marginal cost (MC) is positive, and (3) marginal cost increases as the stock is depleted. New discoveries may occur and thus MC may not rise, but the point is that producers must have some estimate of such events and eventually MC must rise.

The owner of any resource, say oil, will manage his stock so as to maximize the discounted present value (DPV) of the time stream of income from his holdings of the stock:

$$\text{DPV} = \frac{P_1 q_1 - C(q_1, R_1)}{1 + r} + \frac{P_2 q_2 - C(q_2, R_2)}{(1 + r)^2} - + \ldots + \frac{P_n q_n - C(q_n, R_n)}{(1 + r)^n}$$

where P_t and q_t stand for price and output; R_t is the estimated and probable reserves; $C(q_t, R_t)$ are the total costs of mining (not just the extraction costs); and r is the discount rate. The owner of the resource wishes to maximize DPV. This is accomplished by choosing his output rates such that

$$\frac{\Delta \text{DPV}}{\Delta q_t} = 0 \text{ for each } q_t$$

given that for any R_i, increasing output in the tth period reduces the amount that could be mined in the future and assuming that $q_t > 0$. This condition is equivalent to the static rule of choosing output q such that $\Delta \pi / \Delta q = 0$.

An increasing Δq_t in the amount mined in period t increases the net cash flow in the same period by the amount

$$(P_t - MC_t) \Delta q_t$$

But this increase in output reduces the stock available to be mined in period $t + 1$ by Δq_t. Assuming no change in mining rates in periods $t + 2$, $t + 3, \ldots t + n$, the net cash flow in period $t + 1$ is reduced by

$$\left(P_{t+1} - MC_{t+1} + \frac{\Delta C}{\Delta R_{t+1}}\right)\Delta q_t$$

The overall effect on DPV of an increase in output by Δq_t is then given by

$$\frac{(P_t - MC_t)\Delta q_t}{(1 + r)^t} = \frac{\left(P_{t+1} - MC_{t+1} + \frac{\Delta C}{\Delta R_{t+1}}\right)\Delta q_t}{(1 + r)^{t+1}}$$

Thus the rule $\Delta DPV/\Delta q_t = 0$ implies that

$$(P_t - MC_t)(1 + r) = P_{t+1} - MC_{t+1} - \frac{\Delta C}{\Delta R_{t+1}}$$

or

$$P_{t+1} - P_t(1 + r) = MC_{t+1} - MC_t(1 + r) - \frac{\Delta C}{\Delta R_{t+1}}$$

This last expression has a meaningful economic interpretation. Suppose there are no variable costs, and so the right-hand side of the expression is zero. Then, if the resource owner supplies positive amounts of the resource in each period, the price per unit must rise over time at a rate equal to the market rate of interest. That makes a lot of sense. For if $P_{t+1} < P_t(1 + r)$, then the owner should sell his entire stock in period t and invest the proceeds in assets earning the market interest rate.

If, on the other hand, $P_{t+1} > P_t(1 + r)$, then he should sell nothing in period t since he can earn more than the market rate of interest by leaving his stocks in the ground and making the sale in period $t + 1$.

The fact that variable costs are positive complicates things, but the basic idea holds. Before we try to apply this general principle, let us examine what we meant by 'total costs of mining'. In general, mining costs consist of two components: (1) extraction costs or operating costs, and (2) user cost.

Extraction costs are the familiar costs of production. User cost is the increase in cost which is caused by the reduction of the total stock of the resource. That is, the total stock of resources remaining affects the costs of mining. And, since production typically takes place over many periods of time, user cost is determined by the whole future path of costs and prices and not just by current conditions.

In the case of oil, marginal production costs (MC) are close to zero, but user cost, $\Delta C/\Delta R_{t+1}$, is positive. This means that positive output in each period implies that price must rise faster than the rate of interest, since

$$P_{t+1} - P_t(1 + r) = -\frac{\Delta C}{\Delta R_{t+1}}$$

when

$$MC_t = MC_{t+1} = 0$$

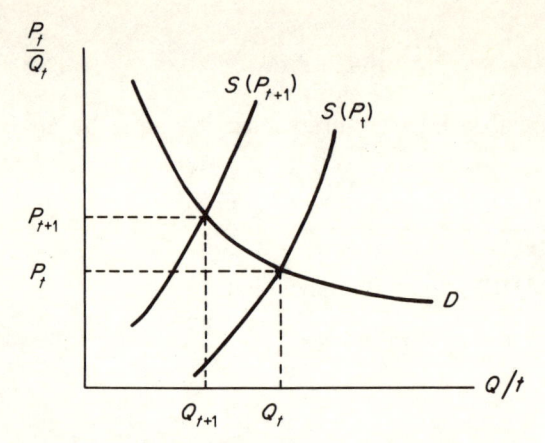

Figure 4.1. Price changes over time in the case of an exhaustible resource.

Since each resource owner will follow this rule, the market for oil will be as depicted in Figure 4.1. Assuming that the demand schedule is unchanged over time (in fact it has been increasing), the market equilibrium is as shown. At each period of time the market supply curve $S(P)$ is the sum of the supply curves of individual producers. At $t + 1$, the supply curve $S(P_{t+1})$ lies to the left of the supply curve $S(P_t)$, because output in period t, Q_t, reduces the stock available for mining at the beginning of period $t + 1$ and, hence, increases the cost of mining any given output in future periods such as $t + 1$. The higher the interest rate, the further to the left is the $S(P_{t+1})$ curve.

As a result, the market price rises over time. If it turned out that large and unexpected new discoveries were made, then the price may not rise. In the case of oil—contrary to what is generally being assumed—no major and unexpected discoveries were made since the late 1950s.

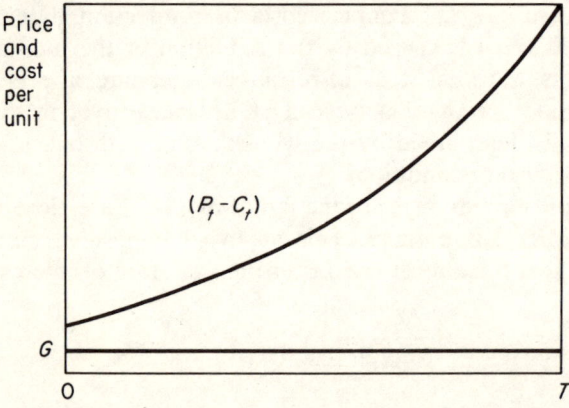

Figure 4.2 The behavior of price over time in the case of constant costs.

For an individual producer, the case of constant cost is illustrated in Figure 4.2. OG is cost per unit, which remains constant over time. The deposit will be exhausted at time T, when the price brings the quantity demanded to zero.

4.2 THE THREAT TO PROPERTY RIGHTS

But how could we reconcile the facts of the oil market with the predictions of this economic model? Firstly, the real posted price of oil between 1947 and 1970 decreased by approximately 65 per cent, even though the interest rate was slowly rising (Adelman, 1977). Secondly, the demand for oil was not constant but rose by about 7 per cent per year. Thirdly, no major and unexpected new discoveries were made since the late 1950s. Was the cost of extraction falling? No, it was not; and the user cost, of course, was rising.

There are two main reasons why in fact the real price of oil fell rather than rose as our analysis predicted.

The first is that the 1947 price of oil was not determined by competitive forces. It was in fact greatly influenced by the Texas Railroad Commission, who instituted demand prorationing to keep prices higher than what they otherwise would have been. Once an oil price was 'posted' at the US Gulf of Mexico, it was then used as the basis of the price of oil throughout the world.

The second reason for the fall of the price of oil has to do with the problem of oil appropriability in the Arabian Gulf. Since the end of World War II, the Arabian Gulf gradually replaced the US Gulf of Mexico as the most important oil-producing area. The governments of the Arabian Gulf signed agreements with the western oil companies to produce oil. But these agreements did not prevent the governments from increasing the oil royalties and threatening the oil companies with nationalization.

Increasing royalties and fear of nationalization increased the oil companies' expected future costs. That led the companies to increase their oil production by a greater rate than if they did not expect future costs to rise. The increase of supply of course led to the fall of prices. A lot more will be said about this in Chapter 5.

The influence of the presence of a threat to the oil companies' property rights on their rate of oil production can be shown formally and more clearly by using the following model.

First let us clarify the notation:

P_t = the unit price of oil in period t
C_t = the average cost of extracting a unit of oil in period t
$R_t = P_t - C_t$
r = the market interest rate

Assume that, as far as the oil companies were concerned, $dc/dt = 0$. Under a world of certain and well-defined property rights, an oil company will supply a positive amount of oil in periods (t) and $(t + 1)$ if it expects

$$R_{t+1} = R_t(1 + r)$$

But the oil companies knew that there was always some threat to their property rights in the OPEC countries. The probability of expropriation varied among countries, but the companies were never one hundred per cent certain that their property rights would not be threatened.

Let us say there is a probability $q < 1$ that nothing will interfere with their operations. Then,

$$E(R_{t+1}) = q(R_{t+1}) + (1 - q)0$$

or

$$\hat{R}_{t+1} = q(R_{t+1})$$

Now they will supply oil in period (t) and period $(t + 1)$ if

$$\hat{R}_{t+1} = R_t(1 + r)$$

That is, R_{t+1} must be greater than $R_t(1 + r)$ in the case of uncertainty. In other words, qR_{t+1} (rather than just R_{t+1}) should be equal to $R_t(1 + r)$, or

$$R_{t+1} = R_t \frac{(1 + r)}{q}$$

The effect of less than 100 per cent certainty about future property rights is to increase the effective rate of discounts, r, on which the oil companies will base their decisions. For example, say the market rate of interest $r = 10\%$, and say a company thinks that the probability q that no

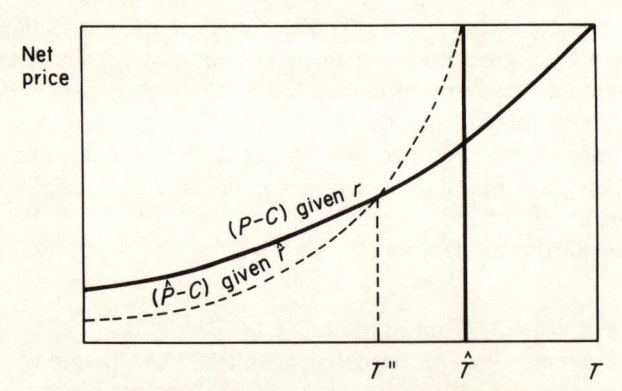

Figure 4.3. The influence of increasing the discount rate on prices over time and on the date of exhaustion.

one will interfere with its operations is 0.75. Then

$$\hat{R}_{t+1} = \frac{R_t(1 + 0.1)}{0.75} = R_t(1 + \hat{r})$$

That is,

$$1 + \hat{r} = \frac{1.1}{0.75} = 1.47$$

or the company's effective discount rate is 47 per cent when $r = 10\%$, $q = 0.75$. The effect of increasing the discount rate on an oil company is obviously to increase its rate of output, which in turn decreases the world's oil price. As long as $q < 1$, however close it may get to 1, the effective discount rate $\hat{r}$ will be greater than the market rate of interest, r. Figure 4.3 shows two possible paths of $(P - C)$, given $\hat{r} > r$.

It is to be expected that uncertainty will result in a greater rate of output. What is surprising is the fact that at $q = 0.75$, a very reasonable approximation, the effective discount rate has changed from 10 per cent to a huge 47 per cent.

REFERENCES

Adelman, M. A. (1977). 'The Changing Structure of Big International Oil' in *Oil, Divestiture, and National Security* (ed. F. N. Trager), pp. 1–11, Crane, Russak, and Company, New York.

Quirk, J. (1976). *Intermediate Microeconomics*, Science Research Associates, Chicago.

CHAPTER 5

OPEC is not a Cartel: an Alternative Explanation

5.1 THE OIL MARKET: 1947–1970

The demand for oil, like the demand for any other commodity, is mainly determined by its price, the prices of its substitutes and complements, and by national income.

In symbolic form,

$$D = F(P_o, P_s, P_c, NI)$$

where P_o is the price of crude oil, P_s and P_c stand for the price of oil substitutes and complements, respectively, and NI is the national income of the oil consuming countries. It is estimated that the income elasticity of demand, η, is somewhere between 1.3 and 1.6. The price elasticity of demand, ε, is thought to be very small in the short run and may range between -0.2 and -0.1.

The important thing about the demand for oil is its sensitivity to changes in the world economies. But in spite of the industrial countries' steady economic growth since the 1940s, the real price of oil has fallen from 1947 to 1970 because the increase in oil supply more than compensated for the increase in demand. Figure 5.1 gives a geometric description of the oil market between 1947 and 1970.

The supply of oil has increased because of continuous oil discoveries, mainly in the Arabian Gulf area. The supply curve has continuously shifted to the right for two reasons: (1) technological progress that reduced the costs of exploitation, and (2) the risks of expropriation that the oil companies feared increased their effective discount rates and hence the rate of their oil outputs.

5.2 THE OIL MARKET: 1970–1973

As we mentioned before, OPEC was created in 1960 mainly to prevent the posted oil price from falling. And as a result, the nominal posted prices

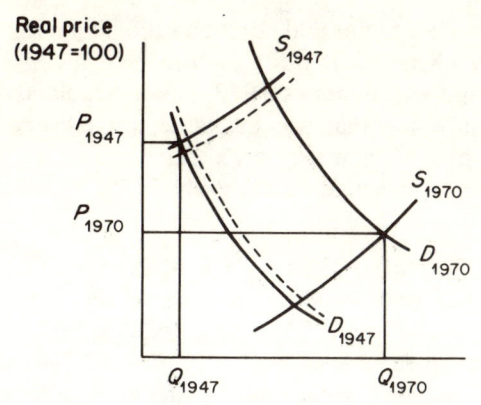

Figure 5.1. The oil market, 1947–1970.

did not fall since the establishment of OPEC, even though many experts believe that the nominal market-determined prices had decreased throughout the 1960s (Mikdashi, 1976: Chapter 2).

Since posted prices and market-determined prices are going to be referred to repeatedly in this chapter, let P_s stand for posted price and P_k denote the market-determined price.

It is important to emphasize the difference between P_s and P_k because, even though P_k is the accurate gauge of the market forces, the figures that are readily available are the published posted prices. All the claims that OPEC is an effective cartel are based on the quadrupling of P_s in the period between January 1973 and January 1974.

Because P_s is used as a reference price to calculate the oil companies' payments to the host countries, it is in the interest of the companies to keep it as low as possible. The reason the companies accepted a P_s that was greater than the P_k in the 1960s was their fear that if they insisted otherwise the countries might increase their income tax rates and their royalties. The companies preferred to accept a higher P_s rather than risk changes in the concessions terms that might in the long run increase their costs by a greater amount than the extra cost that was due to $P_k < P_s$.

When, on the other hand, $P_k > P_s$ it is much easier and politically more acceptable for the companies to bargain for keeping the increases in P_s smaller than the increases in P_k. In other words, during periods of falling P_k we would expect $P_s > P_k$, and during periods of rising P_k we would expect $P_s < P_k$.

What I am trying to establish is the following: If the companies attempted to lower the posted prices, that would have provoked the host countries in such a way as to threaten the companies' entire concessions. Resisting posted price raises, on the other hand, was not as dangerous as attempting to lower them.

The posted price rose sharply since 1971. The important question is if P_s increased by a smaller rate than P_k. OPEC economic experts say most emphatically yes. They claim that, between the beginning of 1971 and June 1973, P_k rose by 65 per cent while P_s rose only by 33 per cent (Kubbah, 1974: 61). During the summer of 1972, Saudi Arabia and Qatar sold all of their participation crude that was available for delivery in 1973 at prices that were 'much higher' than the posted prices.

It seems rather certain that in the early 1970s the %ΔP_k > %ΔP_s. Here is why:

1. Between 1970 and 1973 real GDP of the OECD countries increased by 5.1 per cent per year while the free world oil output increased by 6.6 per cent, and from 1945 to 1960 oil consumption had increased at 150 per cent of the percentage change in GDP; that is, oil output should have increased by 7.7 per cent rather than by the actual 6.6 per cent to keep up with the increase in demand (see Table 5.1 and 5.2).

2. The participation talks between the oil companies and the host countries increased the companies' fear that access to their crude in the OPEC countries would be restricted. That led the 'major' companies to reduce their outside sales of crude, forcing the price of oil that was

Table 5.1. Free world crude oil output in millions of barrels per day

Year	Output	% Change
1970	37.6	—
1971	38.0	0.1
1972	40.7	7.0
1973	45.9	12.7
1974	45.1	−0.2
1975	41.5	−8.0
1976	44.7	8.0
1977	46.8	5.0

Source: *Oil and Gas Journal*.

Table 5.2. Percentage change in gross domestic product of the OECD countries

Years	% Change
1970–71	3.7
1971–72	5.5
1972–73	6.0
1973–74	−0.4
1974–75	−1.5
1975–76	5.3

Source: *Main Economic Indicators*, published monthly by OECD, Paris, France.

available for the third parties to rise. The major companies expected smaller future supplies and thus higher prices and started increasing their inventories; that reduced the total supply of oil which, in turn, increased the prices.

3. In 1970, the US oil output had peaked, and in early 1972 the Bureau of Mines reported that US crude oil output was expected to fall for the third year running by over 2 per cent.

Everyone expected the oil price to rise, and price expectation of course has great influence on actual prices. In the words of Professor Penrose:

> There is a clear record in the trade press of the short-term pressure of demand on market supply and of the increasing prices offered by buyers—mostly American but also Japanese and European independents (Penrose, 1975: 47).

When US oil producers anticipated higher future prices, that led them to reduce their outputs further, and that in turn of course contributed to actual price rise.

5.3 CARTELIZATION OR ALTERATION OF PROPERTY RIGHTS

The *posted* price of OPEC oil was quadrupled in the period between January 1973 and January 1974. I have argued in a previous section that the market-realized price must have increased by a greater rate than the posted price up to the *end of 1972*, and, *therefore*, did not quadruple between the *beginning of 1973 and January 1974*. That is, we are certain P_k had increased, but we are not sure of the magnitude because the data we have only reflect the published P_s.

Regardless of the divergence between P_s and P_k that existed before November 1973, when OPEC for the first time unilaterally determined the oil price, it was clear to everyone that actual P_k sharply rose towards the end of 1973. Furthermore, it is certain that what caused the big price rise is the production cutbacks that were undertaken by OAPEC members of OPEC.

It is true that the output restrictions were carried out for political purposes rather than for the achievement of financial ends. But to an economist that does not matter. What is important is that output reductions made price rises possible.

Since the sharp increase in the market price of oil that followed the October 1973 Arab–Israeli War (although it should be emphasized: the real prices were not quadrupled as has been frequently asserted) economists have assumed that OPEC has become an effective cartel that reduces output to raise prices. I disagree with this explanation.

There is no doubt, of course, that without production cutbacks prices will not rise. I only disagree with the reasons that were advanced to explain the production restrictions.

In the last three months of 1973 there was an oil embargo and there were output reductions, but something else, certainly far more important than all of these things, happened. *The oil producers decided to determine the price of their oil unilaterally rather than through negotiations* with the oil companies as had been done in the past.

Once the host countries became the ones who decided the rate of oil output and its price, the role of the companies was reduced essentially to that of contractors. That amounted to a **de facto** nationalization of the crude-oil deposits.

To outsiders no obvious institutional change took place, because no one announced any change in the contracts that govern the relations between the companies and the host countries.

But, if an oil company could not determine how much to produce and could not even announce the price at which it would sell the oil that it produces, does that make it an owner of that oil? Obviously not.

It is granted, however, that the countries did have the right to receive payments for the extracted oil. The point is that the oil-exporting countries could *not* change the 'posted' price unilaterally, since any price changes had to be negotiated. And until the early 1970s, the companies had considerable influence in determining the final negotiated prices.

Once one recognizes that the ownership of crude oil deposits has been shifted from the foreign companies to the oil-producing countries, it becomes then obvious that the companies and their host countries have different discount rates. And that implies different rates of output, which, in turn, means different levels of prices. A formal simple model was introduced in the previous chapter to show the effect of uncertain property rights on discount rates.

But that is not all. The rise in prices which, it has been argued in this book, was due to output cutbacks which were dictated by wealth-maximizing behavior on the part of the oil-producing countries, also led to further decreases in the countries' discounting rates. This was because the increase in oil revenues dictated reduction in output because of the limited investment opportunities within some of the OPEC countries and the risks of investing abroad for all of the OPEC members.

Aside from all efforts to keep the oil price high, countries such as Saudi Arabia, Kuwait, Venezuela, UAE, Qatar, and Libya had to decrease the rate at which oil was produced for purely efficient resource-allocation purposes. As will be explained in other parts of this book, the marginal rate of return on domestic investment in the Gulf countries is much lower than the average world market rate of interest.

Because the OPEC countries have lower discount rates than the companies' effective discount rate, their oil output since 1973 is lower than what it would have been if the companies were still the owners of the crude. That led to higher oil prices. But the higher prices increased revenues, which in turn decreased the discount rate even further in some OPEC countries, and that resulted in greater production cutbacks.

From Table 5.3 it is clear that since 1973 the supply of oil by Kuwait, Qatar, and Libya has been restricted. One would expect the same thing to occur in Saudi Arabia and in the UAE. But for non-financial reasons that will be explained in Chapter 7, Saudi Arabia and UAE did not restrict their outputs as much as other countries that have similar constraints on domestic investment. It is nevertheless obvious that the rate of increase at which oil was produced in Saudi Arabia was decreased from 25 per cent per year over the 1970–1973 period to only 13 per cent from 1973 to 1974.

Figure 5.2 illustrates what happened since 1973. The demand that OPEC faces, D^e, is the difference between the world total demand for oil and non-OPEC supply. The position of the supply curve reflects two different discount rates, r and $\hat{r}$, where $\hat{r}$ is greater than r.

The sharp increase in the price of oil since 1973 and the slight restriction of oil output that made it possible does not contradict the monopoly power that is being attributed to the allegedly successful cartelization of OPEC.

I am only arguing that the huge difference in the effective discount rates between the foreign companies on the one hand and the host countries on the other is a more reasonable cause of the price rise than the monopoly explanation.

Aside from the political differences among the OPEC countries, the economic differences alone will preclude agreements on prices and rates of output. The economic disagreements could be solved if side payments and demand prorationing were effectuated. But no such thing occurs.

Many economists think OPEC is an effective cartel for two reasons: (1) the sharp price rise since January 1974, and (2) the fact that the OPEC members meet to 'fix' prices.

The price rise could be explained by the change of crude ownership, the boom in the world economies between 1970 and 1973, and the fall of the US crude-oil output.

When it comes to price-fixing, OPEC is not unique. The producers of coffee, tin, copper, and other raw materials meet all the time to fix the prices of their commodities. The problem is that they do not adhere to the agreed-upon prices.

In fact, many, if not all, of the OPEC countries charge different prices than the 'official' price that OPEC announces (Seymour, 1975).

The only time when any number of the OPEC oil producers agreed to act collectively in concert to reduce output was in the last three months of 1973, following the October 5, 1973 Arab–Israeli War.

On November 5, 1973, the OAPEC members of OPEC declared that they would cut their oil outputs by 25 per cent of their September 1973 levels and 5 per cent each successive month until they realized their political objectives. It is important to note that the non-Arab members of

Table 5.3. International crude oil production

		Algeria	Iraq	Kuwait	Libya	Qatar	Saudi Arabia	United Arab Emirates	Arab OPEC	Indonesia
		\multicolumn{9}{c}{Thousand barrels per day}
1973	Average	1,070	2,018	3,020	2,175	570	7,596	1,533	17,982	1,339
1974	Average	960	1,971	2,546	1,521	518	8,480	1,679	17,675	1,375
1975	Average	960	2,262	2,084	1,480	438	7,075	1,664	15,963	1,307
1976	Average	980	2,415	2,145	1,933	497	8,577	1,936	18,483	1,504
1977	Average	1,095	2,495	1,970	2,065	445	9,200	2,000	19,270	1,684
1978	January	1,100	2,130	1,720	1,790	450	7,790	1,740	16,720	1,700
	February	1,100	2,430	1,720	1,800	480	8,380	1,880	17,790	1,700
	March	1,100	2,230	2,130	1,880	420	7,690	1,850	17,300	1,710
	April	1,100	2,430	1,990	1,870	510	8,050	1,750	17,700	1,680
	May	1,100	2,130	1,813	1,930	380	7,250	1,870	16,473	1,700
	June	1,100	2,230	1,925	2,000	450	7,590	1,840	17,135	1,620
	July	1,100	2,100	1,952	2,040	490	7,410	1,830	16,922	1,580
	August	1,100	2,300	2,360	2,030	540	7,180	1,830	17,340	1,620
	September	1,100	3,000	2,591	2,020	500	8,380	1,830	19,421	1,590
	October	1,100	2,700	2,110	2,070	510	9,310	1,840	19,640	1,590
	November	1,100	3,300	2,650	2,100	470	10,250	1,840	20,710	1,590
	December	1,100	3,000	2,199	2,090	580	10,400	1,830	21,199	1,600
	Average	1,100	2,515	2,095	1,975	480	8,295	1,831	18,291	1,635
1979	January	1,100	3,500	2,615	2,175	550	9,790	1,835	21,565	1,605
	February	1,100	3,500	2,705	2,160	555	9,780	1,830	21,630	1,620
	March	1,100	3,500	2,590	2,080	370	9,780	1,825	21,245	1,630
	April	1,100	3,500	2,545	2,070	550	8,790	1,750	20,305	1,610
	May	1,100	3,500	2,585	2,050	540	8,780	1,855	20,410	1,570
	June	1,100	3,500	2,585	2.020	455	8,780	1,865	20,305	1,615
	July	900	3,300	2,550	2,080	520	9,780	1,830	20,960	1,605
	August	900	3,300	2,525	1,990	535	9,770	1,830	20,850	1,600
	September	900	3,300	2,375	2,030	455	9,780	1,835	20,675	1,580
	October	900	3,300	2,375	2,030	490	9,725	1,780	20,600	1,575
	November	900	3,700	2,445	2,095	525	9,795	1,865	21,325	1,575
	December						9,700			
	Average						9,550			

Source: US Department of Energy, *Monthly Energy Review*, February 1980.

OPEC did not participate in the output reductions. To the contrary, oil outputs in Iran, Indonesia, and Nigeria rose during this period.

The actual reductions were far less than the declared targets (see Table 5.4), and the entire OPEC output for the three months of October, November, and December on the average fell by about 8 per cent from the production level of September 1973. It is almost certain that, without the emotional atmosphere that was associated with the war, any collective efforts to reduce outputs would have failed.

One may wonder why the OPEC oil producers waited *until 1973* to be the ones to make the output and pricing decisions.

The answer is two-fold: firstly, the emergence of the so-called for major petroleum exporting countries

Iran	Nigeria	Venezuela	Total OPEC	Canada	Mexico	United Kingdom	United States	China	USSR	Other	World
					Thousand barrels per day						
5,860	2,054	3,366	30,961	1,800	450	8	9,208	1,090	8,420	3,843	55,780
6,022	2,255	2,976	30,683	1,695	580	9	8,775	1,310	9,020	3,799	55,870
5,350	1,783	2,346	27,134	1,420	720	20	8,375	1,490	9,630	4,201	52,990
5,863	2,067	2,294	30,641	1,300	800	245	8,132	1,670	10,170	4,372	57,330
5,665	2,085	2,240	31,350	1,320	980	770	8,245	1,805	10,700	4,490	59,660
5,290	1,640	1,780	27,530	1,240	1,100	880	8,360	1,990	10,900	4,420	56,520
5,530	1,570	1,620	28,600	1,310	1,100	950	8,377	1,990	11,000	4,493	57,820
5,600	1,520	2,060	28,600	1,320	1,100	870	8,720	1,990	11,070	4,620	58,290
5,610	1,690	2,230	29,330	1,100	1,140	980	8,818	1,990	11,100	4,562	59,020
5,720	1,720	2,220	28,253	1,160	1,150	1,110	8,825	1,990	11,140	4,392	58,020
5,630	1,890	2,320	29,015	1,500	1,170	1,110	8,832	1,990	11,120	4,573	59,310
5,800	1,910	2,290	28,952	1,180	1,200	1,090	8,756	1,909	11,230	4,642	59,040
5,810	2,060	2,100	29,330	1,310	1,240	1,100	8,758	1,990	11,280	4,832	59,840
6,050	2,120	2,270	31,881	1,200	1,280	1,090	8,800	1,990	11,340	4,219	61,800
5,490	2,110	2,260	31,520	1,390	1,300	1,160	8,820	2,010	11,440	4,650	62,290
3,490	2,280	2,320	30,840	1,520	1,320	1,280	8,741	1,010	11,490	5,719	62,920
2,370	2,380	2,320	30,299	1,540	1,370	1,350	8,662	2,010	11,470	4,949	61,650
5,200	1,910	2,165	29,616	1,315	1,215	1,080	8,707	2,005	11,220	4,772	59,930
410	2,440	2,270	28,745	1,455	1,390	1,460	8,457	2,280	11,370	4,443	59,600
760	2,430	2,350	29,245	1,580	1,395	1,500	8,498	2,280	11,370	4,322	60,190
2,190	2,440	2,430	30,380	1,410	1,305	1,330	8,585	2,280	11,370	4,390	61,590
3,800	2,420	2,390	30,960	1,515	1,395	1,455	8,533	2,280	11,510	4,508	62,230
4,100	2,400	2,390	31,310	1,470	1,400	1,640	8,585	2,290	11,110	4,395	62,190
3,950	2,420	2,250	30,980	1,470	1,435	1,740	8,409	2,280	11,460	4,466	62,240
3,750	2,380	2,330	31,380	1,525	1,435	1,705	8,355	2,130	11,400	5,480	63,410
3,600	2,185	2,330	30,995	1,455	1,455	1,635	8,699	2,130	11,560	5,250	63,050
3,600	2,115	2,370	30,760	1,495	1,470	1,670	8,466	2,130	11,460	4,979	62,430
3,930	2,135	2,375	31,035	1,450	1,510	1,610	8,460	2.130	11,630	5,120	62,945
3,300	2,150	2,395	31,165	1,530	1,615	1,515	8,530	2,130	11,920	5,000	63,405
			30,800								

newcomers (**Phillips, AMOCO, Oxy,** etc.) and the state-owned oil companies of France, Italy, and Japan as important buyers of crude reduced the monopsony power of the 'majors'.

Secondly, the tightness of the oil market in the 1970–73 period that was caused by actual demand being greater than projected demand reduced the chances that the oil companies would unite and boycott OPEC oil.

In 1954, the oil companies that were nationalized in Iran were able to enlist the support of the other members of the 'Seven Sisters' in blacklisting the Iranian crude. As a result, the Mossadagh government that nationalized the oil industry fell and the companies' property rights were restored.

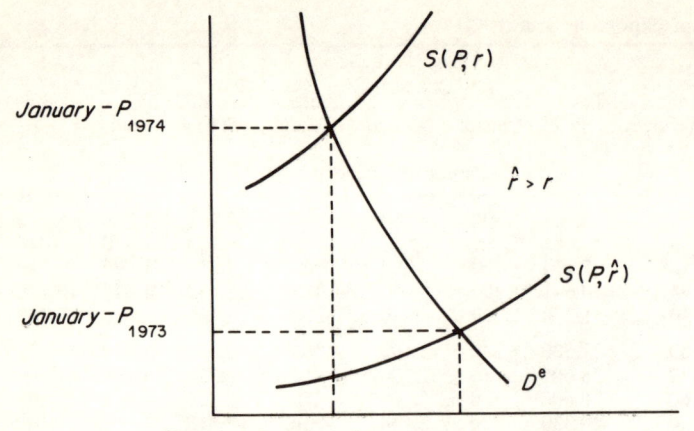

Figure 5.2. The oil market, 1973–1974.

During early 1970 oil buyers were too numerous and had substantially too divergent interests to be able to frustrate the oil producers governments' attempts to control their own natural resources. In short, the power of the host countries was greatly enhanced by the gradual weakening of the economic and political power of the major oil companies.

Table 5.4. OAPEC members' of OPEC production levels during the 3-month 'embargo' period October through December 1973 (as compared to their output levels during September 1973, the month immediately preceding the embargo) (in thousands of barrels per day)

	October	November	December	3-month average	September	Actual % change from September	Planned % change from September
UAE	1,550	1,300	1,160	1,337	1,654	−19	−15
Iraq	1,797	1,923	2,140	1,953	2,167	−9	−35
Kuwait	3,058	2,615	2,560	2,744	3,480	−21	−35
Qatar	598	505	460	521	608	−14	−35
Saudi Arabia	7,800	6,270	6,620	6,897	8,534	−19	−35
Algeria	1,020	800	860	920	1,100	−16	−25
Libya	2,380	1,776	1,770	1,972	2,286	−13	−35
TOTAL	18,203	15,259	15,570	16,344	19,822	−17	−35
Ecuador	210	210	230	217	210		
Venezuela	3,371	3,384	3,330	3,362	3,395		
Iran	5,978	6,009	6,071	6,019	5,393		
Indonesia	1,406	1,391	1,400	1,399	1,350		
Gabon	160	160	160	160	155		
Nigeria	2,190	2,200	2,252	2,214	2,102		
				13,371	12,605	+6	
OPEC Total				29,715	32,427	−8.3	

Source: *Oil and Gas Journal.*

Professor Walter J. Mead (1979) tried to test the hypothesis that OPEC had become a successful cartel since October 1973, and he could not confirm it.

Firstly, if OPEC was really an effective cartel, one would expect the members of the cartel to reduce output 'during market weakness' to maintain what they consider an optimal price. Did they? If we compare the records of output during 1974, 'a year of strong market', with that of 1977, 'a year of weak market', we find that six OPEC members producing approximately one-half of total OPEC output expanded production such that their market shares increased from 52.2 per cent to 58.8 per cent of OPEC production (Mead, 1979: 219). That evidence is not what one would expect if OPEC was an effective cartel. See Table 5.5.

Secondly, Professor Mead looked at the evidence (Table 5.6) along Robert Pindyk's classification of the OPEC members as 'saver countries' and 'spender countries'. If collusion was happening then one would expect 'saver' countries to be the members of the 'cartel' who were bearing the brunt of output reduction. 'The record, however, shows that "saver countries" in total expanded output and market shares at the expense of the spender countries' (Mead, 1979: 219).

5.4 BUT DOES NOT SAUDI ARABIA DOMINATE OPEC?

It is of great importance to note that saying that 'OPEC is a cartel' is quite different from saying that 'Saudi Arabia determines the price of oil'. In other words, it is possible that OPEC is not a cartel even if one thinks that Saudi Arabia has a monopoly power in the world oil market.

It has been often said that OPEC is a cartel and Saudi Arabia is its price leader. But that does not make sense. If OPEC is a cartel then its oil price is the price that has been set to serve the interest of *all* the members of the cartel. If, on the other hand, Saudi Arabia is able to set the OPEC price then that price is supposed to serve mainly the interest of Saudi Arabia. Under the rules of the dominant-firm price-leadership, Saudi Arabia would set what it considers an optimal price, and all the other oil-producing countries would sell as much as they want at that price.

But does the behavior of Saudi Arabia and the other oil-producing countries confirm the predictions of the price-leadership model?

From January 1974 to December 1977, Saudi Arabia seemed to have some power in influencing the world oil price. It did not use this power, however, to increase prices; it used it rather to avert price rises, or at least it tried to keep the price rises as small as possible. Since the demand for oil is considered highly inelastic, and since the marginal cost of producing oil in Saudi Arabia is considered constant, one would expect Saudi Arabia (if it is trying to prevent the price from falling) to reduce output during periods of market weakness while allowing all the others to expand it if they wished. The records show (see Table 5.5), however, that Saudi Arabia

Table 5.5. OPEC country output and market shares, classified by expanding and contracting countries

OPEC Member Country	1974 Output per day (000 bbls)	1974 Market share in OPEC (per cent)	1977 Output per day (000 bbls)	1977 Market share in OPEC (per cent)
Countries expanding output:				
Saudi Arabia	8,481	27.6	9,200	29.5
Iraq	1,975	6.4	2,265	7.3
Libya	1,521	4.9	2,080	6.7
Indonesia	1,375	4.5	1,685	5.4
United Arab Emirates	1,689	5.5	2,009	6.4
Algeria	1,009	3.3	1,123	3.6
Total	16,050	52.2	18,362	58.8
Countries contracting output:				
Kuwait	2,547	8.3	1,969	6.3
Iran	6,022	19.6	5,699	18.3
Venezuela	2,976	9.7	2,238	7.2
Nigeria	2,256	7.3	2,097	6.7
Others (3)	895	2.9	849	2.7
Total	14,696	47.8	12,852	41.2
Total OPEC	30,746	100.0	31,215	100.0
Total world	56,268		59,798	
OPEC share of world output		54.6		52.2

Reproduced by permission of *The Journal of Energy and Development*.

increased its absolute output as well as its market share, 'both within OPEC and in total world market' (Mead, 1979: 221).

Until December 1977 Saudi Arabia had a lot of power to influence prices because its capacity to produce oil was estimated to be around 12 million b/d. Thus it was able to argue within OPEC for smaller or no price rises, and it increased its rate of output when necessary to prevent price rises. No one within OPEC doubted, *at that time*, that Saudi Arabia would increase its output to the level which made price rises impossible.

Throughout 1976 most OPEC members wanted to raise the price of oil from its 1975 level, but Saudi Arabia refused and increased its rate of production during 1976 by 21 per cent over 1975 to make its refusal of price increases effective.

When the Iranian revolution resulted in oil output cutbacks during the fall of 1978 the price of oil rose sharply in the spot market, and Saudi

Table 5.6. OPEC country output and market shares, classified by 'saver countries' and 'spender countries'

	1974 Output per day (000 bbls)	1974 Market share in OPEC (per cent)	1977 Output per day (000 bbls)	1977 Market share in OPEC (per cent)
'Saver countries'				
Saudi Arabia	8,481	27.6	9,200	29.5
Kuwait	2,547	8.3	1,969	6.3
Libya	1,521	4.9	2,080	6.7
United Arab Emirates	1,689	5.5	2,009	6.4
Total	14,238	46.3	15,258	48.9
'Spender countries'				
Iran	6,022	19.6	5,699	18.3
Venezuela	2,976	9.7	2,238	7.2
Algeria	1,009	3.3	1,123	3.6
Indonesia	1,375	4.5	1,685	5.4
Nigeria	2,256	7.3	2,097	6.7
Ecuador	174	0.6	183	0.6
Total	13,812	44.9	13,025	41.7
Unclassified				
Iraq	1,975	6.4	2,265	7.3
Gabon	202	0.7	222	0.7
Qatar	519	1.7	445	1.4
Total	2,696	8.8	2,932	9.4
Total OPEC	30,746	100.0	31,215	100.0

Reproduced by permission of *The Journal of Energy and Development*.

Arabia raised its oil output to 9,310 million b/d in October and to over 10,000 million b/d for the months of November and December 1978.

Throughout 1979 Saudi Arabia raised its oil output from the government's announced target of 8.5 million b/d to 9.5 million b/d. Obviously that was done to avoid price rises.

But Saudi Arabia lost control of oil prices in mid-1979. The reason seems to be that it could not increase its rate of output beyond 10 million b/d without causing serious damage to its oil fields (see Chapter 8, section 8.4). The US Central Intelligence Agency in 1979 estimated Saudi Arabia productive capacity to be around 10 million b/d instead of the 12 million b/d that was previously announced.

The behavior of prices since the start of the Iranian political turmoil will be discussed in greater detail in the following chapter.

5.5 AN ALTERNATIVE HYPOTHESIS: THE COMPANIES AS TAX COLLECTORS

It has been argued in this book that since 1974 the presence or absence of OPEC as an organization has not really mattered as far as the oil prices were concerned.

Other writers have thought that OPEC, as an organization, does for the oil producers what a joint sales agency does for a centralized cartel. That is just not the case.

Professor M. A. Adelman advanced another explanation that is based on an institutional factor which is peculiar to the member countries of OPEC. According to Adelman, when OPEC sets its price at a certain level it does so by raising its member governments' taxes per barrel of oil. These taxes, that have to be paid by the producing western oil companies, 'are in the form of income taxes, in fact excise taxes, in cents per barrel. Like any other excise tax, they are treated as a cost and become a floor to price'. If it was not for this OPEC tax system, the 'cartel would crumble' because,

> The floor to price would then be not the tax-plus cost, but only bare cost. The producing nations would need to set and obey production quotas. Otherwise, they would inevitably chisel and bring prices down by selling incremental amounts at discount prices. Each seller nation would be forced to chisel to retain markets because it could no longer be assured of the collaboration of all other sellers. Every cartel has in time been destroyed by one then some members chiselling and cheating; without the instrument of the multinational companies and the cooperation of the consuming countries OPEC would be an ordinary cartel (Adelman, 1973: 87).

The main flaw of Adelman's analysis is this: Why does not any member of OPEC that wants to increase its share of the market just simply reduce its tax rate by 'an incremental amount'?

Since the alleged OPEC cartel price is so high, an individual member country must face a highly elastic demand curve and can increase its revenues by a small tax reduction. Granted that an oil company could not reduce the price at which it will sell oil if OPEC taxes account for almost the entire price, but an oil producing sovereign nation certainly can reduce its tax rate to increase its market share.

REFERENCES

Adelman, M. A. (Winter 1973). 'Is the Oil Shortage Real? Oil Companies as OPEC Tax Collectors', *Foreign Policy*, 69–107.

Kubbah, A. (1974). *OPEC: Past and Present*, Petroleum Economic Research Center, Vienna.

Mead, W. J. (Spring 1979). 'An Economic Analysis of Crude Oil Price Behavior in the 1970s', *The Journal of Energy and Development*, 212–228.

Mikdashi, Z. M. (1976). *The International Politics of Natural Resources*, Cornell University Press, Ithaca, New York.

Seymour, I. (June 20, 1975), 'Iraqi Oil Policy in Focus', *Middle East Economic Survey*, 1–6.

CHAPTER 6

The Iranian Revolution and the Oil Price Explosion

6.1 TEMPORARY AND PERMANENT EFFECTS

The immediate effect of the Iranian revolution was to increase buyers' uncertainty about future oil supplies. And that led to an increase in demand by the oil companies, who tried to build up their inventories, as well as to an increase in the purchases by speculators who expected prices to rise. The net result was that prices did in fact rise even before the Iranian output cutbacks were actually carried out.

Yet, the effect of the Iranian political changes on prices in the long run is not going to be significant. The reason is that oil is a non-renewable resource, and as a result output reductions now simply mean more oil will be available in the future. And output increases by other countries, such as Saudi Arabia, to offset the short-run effects of the Iranian output reductions implies less oil in the countries that increased output could be extracted in the future.

The net effect, therefore, of the Iranian political turmoil, in the long run, as far as prices are concerned, is not as important as may first appear.

It is obvious, but nevertheless worth repeating, that politics can affect the price of any commodity (yes, including oil) only if it influences the rates of supply and/or demand both in the present as well as in the future. Thus, the Iranian revolution led to oil price rises because it affected the current and the expected supply of oil. But oil, unlike other reproducible commodities, is different in the sense that output that is lost now is simply available for the future. This is not the case if the commodity is reproducible. For example, if total world output of rice was reduced due to some political upheavals, that does not mean more rice would be available in the future; that is, the lost output of a reproducible commodity is lost forever (Mead, 1979).

The most *permanent* effect of the Iranian revolution on the oil market is, however, the increase of the relative importance of the *spot market*. Because prices were changing so fast during 1979 oil producers everywhere started announcing the level of prices based on information revealed by the spot market, which reacts to changes in the world supply of and demand for oil much faster than the OPEC organization.

It is estimated by trade journals that 15–20 per cent of the world oil output was sold in the spot market in November 1979, as compared to only 3–5 per cent in January 1979. Markets are social inventions that perform a very useful function—they reduce the costs of exchange. But oil, for reasons that were explained in detail in Chapter 2, was exchanged chiefly through long-run contracts between either the major oil companies and refiners or between the oil-producing countries and the major oil companies.

Recently, however, due in large part to the effects of the Iranian oil output cutbacks, everyone started dealing in the spot market, including the major oil companies themselves and many of the oil-producing countries (*Petroleum Economist*, January 1980: 2).

6.2 THE EVOLUTION OF PRICES, JANUARY 1978–JUNE 1979

From the middle of 1977 to the end of 1978 the nominal price of oil was kept at $12.70 for a barrel of 34° Arabian Light (OPEC's marker crude), even though the real price of oil was falling continuously over the same period due to the decline in the purchasing power of the dollar.

Of course, the oil exporting countries wanted to raise the nominal price of oil to compensate them for the fall in the real value of the dollars they received as payments for their oil. The price of oil, however, is not determined by the hopes of sellers or the wishes of buyers, it is simply determined by the world's total supply and total demand. And during 1978 OPEC's total supply fell slightly from its level of 1977, which reached 31,350 million b/d, to 29,616 million b/d, mainly because it was not possible to sell more than that at the prevailing prices—at least up to November 1978.

The political upheavals in Iran, which started in the spring of 1978, had finally affected Iranian oil output in the autumn of 1978. The daily average rate of output fell from 5,490 million b/d in October to 3,490 million b/d in November, and by December it reached 2,370 million b/d.

The effect of reducing oil output in Iran of course was an increase in oil prices in the spot market. But because the reason behind output reductions in Iran was political unrest, the increase in spot market prices was more than what one would expect purely as a result of a decrease in oil supplies. That is, expectations about the future were more powerful than actual events at that time.

Thus, when the 13 ministers from OPEC countries met in Abu Dhabi in December 1978, it was a foregone conclusion that they would raise the price of 34° Arabian Light from $12.70, the price which had prevailed for the previous 18 months. The only uncertainty was about the magnitude of the price increase.

The ministers agreed to increase the prices by an average of just 10% for the entire year of 1979. They were to rise in four quarterly instalments.

But the Iranian 'crises', which made the December 1978 price increase possible, rendered them completely useless by February 1979. How? At the beginning of 1979 Iranian oil exports ceased completely. That led to actual price levels that were far higher than the agreed-upon prices during the December 1978 meeting in Abu Dhabi.

On March 1979, UAE and Qatar increased their sales prices by about a dollar a barrel in addition to the previously announced quarterly increases. Kuwait and Venezuela imposed a surcharge of about $1.20 a barrel. The African producers also raised their prices by even greater amounts.

Why did these countries raise their prices by a greater percentage than was announced in the OPEC meeting of December 1978?

For one reason only: these countries found that what was demanded was greater than what they in fact intended to sell. In such a case market equilibrium (equality of supply and demand) requires an increase in output and/or an increase in price. Saudi Arabia for its own reasons opted for output increase; the other countries chose a price rise.

It turned out, however, that the increase of Saudi output and the price increases that were announced by oil producers throughout the world were not sufficient to equate oil demand to oil supply. With the passage of time oil buyers became convinced that what was happening in Iran was in fact more than just political unrest. Thus they increased their oil stocks. Speculators expected even higher prices and increased their purchases of oil. The net result was a continuous increase in demand during the entire year of 1979.

By the end of May 1979 spot prices reached $38 a barrel. It was obvious that the 'official' price served only the interest of the major oil companies since the actual consumer would have to pay the marginal price (the real cost of obtaining more oil), which was around $35, and the 'majors' had to pay only the 'official' price, which was less than $15.

As a result, when the OPEC delegations met in Geneva during June 1979 again a price rise was certain. But a unanimous agreement on magnitude of price rises could not be achieved.

Saudi Arabia announced reluctantly a price increase of $3.454 per barrel—from $14.546 to $18. Other OPEC members announced greater price advances. It was also agreed that prices should not exceed $23.50.

6.3 THE EXPLOSION OF PRICES, NOVEMBER 1979–FEBRUARY 1980

It was thought that the world oil price would stabilize in the second half of 1979 after the members of OPEC had met in Geneva toward the end of June and raised their official prices. And for a while that happened.

By the end of July the price of Arabian Light, the marker crude, had fallen to around $31 a barrel from $34/$35 a month earlier. During the first week of August the gap between spot and official prices was only $10 as compared to $20 at the end of May (*Petroleum Economist*, August 1979: 346).

The main reason for the relative stability of the market during July and August was an increase in the world total supply following the official price rises that were announced in the last week of June 1979.

But during the fall of 1979 spot prices started rising again as buyers' uncertainty about future supplies increased due mainly to continuous political upheavals in Iran. Toward the end of September the spot price of Arabian Light was $35–36 per barrel, $5 to $6 above its level at the beginning of August. By October the spot price had climbed to $38 per barrel (*Petroleum Economist*, January 1980: 42).

Many producers became convinced that their official prices again became obsolete and served as a very effective means of increasing the wealth of 'middle men' who paid official prices and sold in the spot market at much higher prices. Even the major oil companies could not resist the temptation of the spot market and started selling oil in it.

As a reaction to higher prices in the spot market many of the producers raised their official selling prices. By October 20, Libya announced $26.27 as the new price of its oil, Iran raised it to $23.71. Both prices were above OPEC ceiling of $23.50 (*Saudi Business*, December 21, 1979).

Mexico, a non-OPEC member raised the price of its oil from $22.60 to $24.60. China was able to get $24 for each barrel of oil it sold to Japan.

Peru sold 800,000 barrels on the spot market at $40.77 a barrel.

By December 1979 34° Arab Light had reached $45 a barrel in the spot market.

No one was adhering to the 'official' price of $18 except Saudi Arabia.

'In desperate attempt to regain its lost initiative, Saudi Arabia announced in mid-December that it would raise the marker by $6', which meant the official Saudi price of Arabian Light had reached $24 a barrel (*Petroleum Economist*, January 1980: 2).

Thus during the OPEC conference which took place in Caracas toward the end of December 1979, Saudi Arabia tried very hard to convince all the members of OPEC to declare a $24 a barrel price, with allowance for quality differentials and different costs of transporting crude to markets.

But that was not possible when the oil producers knew that the spot market price of Arabian Light was around $45 a barrel. Thus the conference ended (as happened during the June 1979 conference) with no unanimous agreement on prices.

The invasion of Afghanistan by the Soviet Union (December 27–31, 1979) added to the uncertainty about future supplies, and that of course led to greater demand which resulted in even higher prices.

Thus toward the end of January 1980, Saudi Arabia raised the price of

34° Arabian Light from $24 to $26 hoping to reduce world demand and reduce speculators' 'profits'.

Kuwait, UAE, and Iraq also raised their prices, making the Saudi 'official price' of $26 a barrel the lowest price that buyers could pay for a barrel of oil.

At those high prices, specialists had said, a 'glut' of oil would materialize by the end of the spring of 1980.

6.4 BLAMING THE OIL COMPANIES

The oil companies have always been accused of increasing their own wealth at the expense of consumers and producing countries alike. And perhaps there was a time in the distant past when these accusations were justified on economic grounds.

But the price rises that occurred during 1979 were due to shortages which the oil companies did not initiate. The main cause of the shortage was of course the political changes that occurred in Iran.

When a shortage develops, it can be eliminated only if supply can be increased and/or if demand can be decreased. Saudi Arabia tried to eliminate the shortage of oil by increasing the quantity supplied, and the oil companies tried to eliminate it by decreasing the quantity demanded—thus raising prices.

Even when the real reason for the price rises were events in Iran which could not be affected by either OPEC or the oil companies, OPEC and the oil companies were blamed for them!

OPEC was blamed because consumers were told time and again that this organization is a *cartel* that can increase prices whenever it wants to. The oil companies were blamed when their 'profits' increased by high rates.

In fact the 'profits' of the companies did rise a great deal. The reason: they bought oil at low 'official' prices and had accumulated large inventories before the market prices rose. That is not to say that the oil companies do not want to increase their 'profits' whenever they can, it is only to emphasize that if an oil company has only 8 million barrels for example and refiners wanted to buy 10 million barrels at $20 a barrel, there is only one way to make the quantity demanded equal to the quantity supplied, and that is by raising the price. Unexpected price increases, however, do lead to higher rate of profits.

It should be noted here that the purpose of this section is not to make a moral assessment of any kind; it is rather to explain the objective reasons for the increase of the companies' 'profits'.

REFERENCES

Mead, W. J. (Spring, 1979). 'An Economic Analysis of Crude Oil Price Behavior in the 1970s', *The Journal of Energy and Development*, 212–228.
Petroleum Economist, August 1979.
Petroleum Economist, January 1980.
Saudi Business, December 21, 1979.

CHAPTER 7
Saudi Economic Choices

INTRODUCTION

To Saudi Arabia, a barrel of oil in the ground is a capital asset much like a computer or a cement plant. The only difference is that oil is not reproducible, so the stock of known reserves, given a state of technology and an oil price, could not be increased without further efforts (investment).

The only way the stocks of oil in the ground can generate income is by appreciating in value. The country's assets should be managed in such a way that all assets—oil or buildings—earn the same rate of return, with adjustments for risk. Thus, if oil is to be left in the ground, its net price (price minus operating and user costs) must increase at a rate equal to the interest rate.

The difficulty, as we will explain below, is that Saudi Arabian domestic investment opportunities in the short run are limited, and there are no riskless foreign assets in which it can invest. Furthermore, it has non-pecuniary objectives that also blur an already unclear picture.

The reasons for the low rate of return on domestic investment are not hard to understand once one recognizes the degree of backwardness that characterizes the non-oil sector of the economy. 'In 1940 the wheel was not in general use in most areas of the nation.' 'Saudi Arabia,' wrote an American economist, 'had a pastoral economy based on the raising of goats, sheep, and camels. The majority of the urban population lived in small villages built of mudbrick and earned a living from subsistence agriculture.' (Knauerhase, 1975: 57).

Today the Saudi economy is not as backward as it was in 1940, and the number of 'wheels' in the streets make driving an automobile in Saudi cities as enjoyable (and certainly more dangerous) as driving it in the streets of San Francisco or Tokyo or Paris.

Yet, in many ways the Saudi economy is still backward. And the acute shortage of skilled and unskilled labourers is making the process of development almost impossible. The majority of teachers (even on the elementary school levels), doctors, nurses, engineers, etc., are imported from other countries.

Furthermore, in spite of the shortage (or absence) of professional and skilled workers in Saudi Arabia, the greatest problem that faces anyone who contemplates undertaking investment projects is the lack of information and sufficient public infrastructure. Of course, information is not free in any society, but in Saudi Arabia sometimes it is not available at any price.

An entrepreneur who is considering an investment project in most of the western countries only worries about the costs of his factors of production rather than about their availability. Furthermore, he knows that with some effort he will easily obtain figures on things that greatly affect the cost and/or the price of his product or service.

7.1 ABSORPTIVE CAPACITY

In some less developed countries, such as Saudi Arabia, some factors of production are either not available, or available at costs that make the rate of return on an entrepreneur's capital less than what he can get by depositing his money in a foreign bank. Development economists have coined a term to describe such affairs. The term they came up with is 'absorptive capacity'.

What is absorptive capacity? J. H. Adler defines absorptive capacity as 'that amount of investment . . . that can be made at an acceptable rate of return, with the supply of co-operant factors considered as given' (Adler, 1965). This definition does not specify the 'acceptable rate of return'. But one may say that, in the case of Saudi Arabia, an acceptable rate of return is the rate that exceeds or is equal to the rate that Saudi Arabia can obtain on its financial capital by investing it in the form of foreign bonds with adjustment for risk.

Of course, the Saudi government does invest for the purpose of increasing the supply of the 'co-operant factors'. Nevertheless, in the short run this increase is either physically impossible or it increases the costs of investment in such a manner as greatly to reduce the rate of return on the investment projects in their entirety.

After some careful reading and reflections on the concept of 'absorptive capacity', one may conclude that it essentially refers to the same natural phenomenon that the law of diminishing returns describes.

In terms of absorptive capacity, we may say that Saudi Arabia could not invest all of its financial capital within its borders because the supply of other 'co-operant' factors could not be augmented as easily as the supply of financial capital.

In other words, the marginal product of financial capital falls faster than the marginal product of other factors because financial flows are more abundant relative to other factors. But that is precisely what the law of diminishing returns means. The law of diminishing returns can be stated quite briefly.

Consider a general production function for the entire economy, which specifies that the country's output is a function of the combination of inputs X_1 through X_n. As a result of scarcity, all inputs could not be augmented at the same rate forever. The marginal product of the factor that cannot be increased will eventually rise. This fixed factor is called 'co-operant' factor in the absorptive capacity literature.

The purpose of the above discussion of the law of diminishing returns is not to confuse a well-defined concept, but is to emphasize that what limits Saudi Arabian domestic investment opportunities, and what development economists refer to as its absorptive capacity, is in fact that law of nature, the law of diminishing returns. That is to say, Saudi Arabian expenditure on domestic investment could not be undertaken at the same rate at which the country's financial stocks are increasing, because the supply of other factors could not increase as fast.

What makes the economic experience of Saudi Arabia interesting is the relative abundance of its financial assets. With the exception of a few other OPEC members, one of the major characteristics of less developed countries is the low rate of capital formation. As a result, many economists thought that if capital formation could be greatly increased through domestic savings and foreign aid, then economic backwardness would be defeated.

The Saudi economy, however, does not only empirically confirm the law of diminishing returns, but also shows that the availability of financial capital does not transform an underdeveloped country into an industrial nation within a few years.

The above statements can be briefly formalized. Let us say that the physical relationship between Saudi Arabia's inputs of factors of production and its outputs of goods and services is described by the general form,

$$Q = f(A, R, S, G, Z)$$

where

Q = output per unit of time
A = capital in the form of financial assets
R = infrastructure such as roads, ports, and telecommunications system
S = skilled labor
G = managerial skills
Z = other factors such as financial institutions, health care facilities, population growth, etc.

The supply of factor A is more abundant than any of the other factors, R, S, G, and Z. What the development process usually accomplishes is to increase the MP of A, the relatively abundant factor, by increasing the supply of relatively scarce factors.

What Saudi Arabia wants to accomplish is to make $(\partial Q/\partial A)$ at period $(t + 1)$ greater than $(\partial Q/\partial A)$ at period (t) by increasing the supply of other factors such that the supply of (R, S, G, Z) at period $(t + 1)$ becomes greater than the supply of (R, S, G, Z) at period (t), where $t = 1, 2, 3, \ldots,$ n^{th} future years. Development economists refer to this process as increasing absorptive capacity. We might just as well call it the expansion of productive capacity.

7.2 THE SAUDI REALITIES

Having discussed the general theoretical principles behind the limited investment opportunities in Saudi Arabia, we now turn to the actual figures. Table 7.1 confirms the very well-known fact that the Saudi economy is dominated by the oil sector, which accounts for about 60 per cent of the nation's GDP. The other 40 per cent of the Saudi GDP is chiefly generated through activities created and/or funded by the oil revenues. (The source of information embodied in this chapter is the annual reports of the *Saudi Arabian Monetary Agency*, or SAMA, as it is commonly known.)

Therefore, one would expect the economic activities in Saudi Arabia to be governed by the rates of oil output and level of prices; see Tables 7.2 and 7.3.

Oil revenues rose from \$4,340 million during fiscal year 1973 to \$22,573.5 million during fiscal year 1974. No one of course should expect the Saudi economy's absorptive capacity to quadruple in one year (between December 1973 and December 1974) simply because oil prices rose by more than 400 per cent.

Thus the Saudi government had to increase the economy's absorptive capacity as fast as possible and invest abroad the excess of revenue over expenditure.

From Table 7.4 we can see that the government's total expenditure was about one-third of total revenue in fiscal year 1974/75. The difference between revenue and expenditure for that year (100,103 − 32,038 = 68,065 million riyals) had to be invested abroad. (At that time, one US dollar = 3.53 Rls.)

In the process of trying to adjust the economy for the increased oil revenues two problems emerged: one was an increase in the rate of inflation, and the other was the awareness on the part of Saudis that investing their funds in foreign countries was and is a very risky endeavor. (The last point will be discussed in greater detail in the following section.)

When oil revenues rose government expenditure was also increased. And since the Saudi government finances its expenditure by selling foreign exchange to SAMA (the Saudi equivalent of a central bank) for Saudi riyals, the money supply also rose sharply.

M_3, the broad measure of money supply (particularly useful in Saudi Arabia) increased at an astronomical rate of almost 74 per cent in fiscal year 1974/1975 and at approximately 53 per cent in the following year.

Table 7.1. Gross domestic product by economic activity at current prices

	Fiscal Year					
	1973	1974	1975	1976	1977	1978
	(Million Riyals)					

A—Industries and other products except producers of government services:

	1973	1974	1975	1976	1977	1978
1. Agriculture, forestry, and fishing	1,139 (2.8)	1,242 (1.2)	1,392 (1.0)	1,586 (1.0)	1,866 (0.9)	2,255 (1.0)
2. Mining and quarrying:						
(a) Crude petroleum and natural gas	26,284 (64.8)	73,345 (78.9)	104,696 (75.0)	109,560 (66.6)	128,466 (62.6)	126,156 (56.4)
(b) Other	90 (0.2)	146 (0.1)	264 (0.2)	535 (0.3)	823 (0.4)	1,025 (0.5)
3. Manufacturing:						
(a) Petroleum refining	1,811 (4.5)	4,347 (4.4)	5,766 (4.1)	5,962 (3.6)	6,221 (3.0)	5,908 (2.6)
(b) Other	617 (1.5)	730 (0.7)	1,600 (1.1)	2,211 (1.3)	3,063 (1.4)	4,066 (1.8)
4. Electricity, gas, and water	319 (0.8)	328 (0.3)	195 (0.1)	151 (0.1)	144	204 (0.1)
5. Construction	1,809 (4.5)	2,720 (2.7)	7,719 (5.5)	15,854 (9.6)	25,546 (12.4)	31,959 (14.3)
6. Wholesale and retail trade, restaurants and hotels	1,554 (3.8)	2,355 (2.4)	3,897 (2.8)	6,180 (3.8)	8,507 (4.1)	11,049 (4.9)

7. Transport, storage, and communication	2,121 (5.2)	2,718 (2.7)	2,310 (1.7)	4,077 (2.5)	6,775 (3.3)	9,960 (4.5)
8. Finance, insurance, real estate, and business services:						
(a) Ownership of dwellings	1,000 (2.5)	1,333 (1.3)	3,425 (2.5)	5,278 (3.2)	6,924 (3.3)	7,632 (3.4)
(b) Other	523 (1.3)	746 (0.8)	2,012 (1.4)	3,166 (1.9)	4,206 (2.0)	5,072 (2.3)
9. Community social and personal services	339 (0.8)	403 (0.4)	1,281 (0.9)	1,989 (1.2)	2,609 (1.2)	3,293 (1.5)
10. Less imputed bank service charges	−51 (−0.1)	−64 (−0.1)	−324 (−0.2)	−547 (−0.3)	−928 (−0.4)	−1,561 (−0.7)
Sub-total	37,555 (92.6)	95,349 (96.0)	134,233 (96.1)	156,002 (94.8)	194,222 (94.7)	207,018 (92.6)
B—Producers of government services:						
1. Public administration and defence	1,363 (3.4)	1,858 (1.9)	2,689 (1.9)	3,914 (2.4)	4,998 (2.4)	9,204 (4.1)
2. Other services	1,170 (2.9)	1,632 (1.6)	2,301 (1.7)	3,976 (2.4)	4,722 (2.3)	5,942 (2.7)
Sub-total	2,533 (6.3)	3,490 (3.5)	4,990 (3.6)	7,890 (4.8)	9,720 (4.7)	15,146 (6.7)
GDP in Producer's Values	40,088	98,839	139,223	163,892	203,942	222,164
Import duties	463 (1.1)	475 (0.5)	376 (0.3)	634 (0.4)	1,114 (0.5)	1,583 (0.7)
GDP in Producer's Values	40,551 100.0	99,315 100.0	139,599 100.0	164,526 100.0	205,056 100.0	223,747 100.0

Note: Figures in parenthesis indicate the relative share of each activity in total GDP.
Source: SAMA, *Annual Report*, 1979.

Table 7.2. The price of Saudi Arabian Crude

Year	Posted price
1973	$ 2,591
1974	11,651
1975	11,251
1976	12,376
1977	13,000
1978	13,660
1979 (June)	18,000
1979 (December)	24,000
1980 (February)	26,000

Source: Saudi Arabian Monetary Agency (SAMA), *Annual Reports*, 1974–1979.

It should come as no surprise, therefore, that prices started rising in 1974, given the level of liquidity attained in that year. According to the government figures, the rate of inflation increased from less than 5 per cent in 1972 to 40 per cent in 1976.

The economic policy makers in Saudi Arabia became convinced that the price the country had to pay for increasing the economy's absorptive capacity at such a fast rate was an unacceptable level of inflation.

As a result, the emphasis was shifted from increasing productive capacity to holding down the rate of increase of the general price level. The growth of total expenditure, and thus the growth of the money supply, was slowed down a great deal. Between fiscal years 1977/1978 and 1978/1979 the growth rate of M_3 was only 14.5 per cent, about 1/5 of its growth rate during fiscal year 1974/1975. And that in turn led to a decrease in the inflation rate from 40 per cent during 1974/1975 to only 3.47 per cent between mid-1978 and mid-1979 (see Appendix A to this Chapter).

But the government efforts to increase the economy's productive capacity in the years 1974–1977 did not only increase the rate of inflation but also increased the economy's productive capacity such that in fiscal years 1977/1978 and 1978/1979 total expenditure exceeded total revenue and the deficit had to be financed by the reserve that had been accumulated in previous years (see Table 7.4). In fact Saudi Arabia experienced in 1979 something which it was not used to—a balance of payments deficit of almost $1 billion.

7.3 THE RISKS AND REWARDS OF FOREIGN INVESTMENT

For the world as a whole it is meaningless to ask when we will run out of the last drop of oil, because we will not. But for Saudi Arabia, as an individual country that derives almost all of its foreign exchange from oil

Table 7.3. Oil revenue by source (Million US Dollars)

Year	Aramco	Getty oil	Arabian oil	Other oil companies	Total
1966	745.8	20.6	21.4	2.1	789.9
1967	853.2	17.8	31.8	0.8	903.6
1968	872.0	13.6	34.3	6.5	926.4
1969	895.1	15.2	37.1	1.7	949.2
1970	1,148.4	17.2	40.3	8.1	1,214.0
1971	1,806.4	20.6	44.2	13.7	1,884.9
1972	2,643.2	28.0	68.7	4.7	2,744.6
1973	4,195.0	2.0	91.4	31.7	4,340.0
1974	22,375.0	53.3	113.6	31.6	22,573.5
1975	24,838.6	191.1	642.7	3.8	25,676.2
1976	29,937.3	254.7	559.2	3.6	30,754.9
1977	35,703.8	263.4	571.6	1.2	36,540.1
1978	31,609.0	286.6	338.2	—	32,233.8
1977					
1st Quarter	8,568.3	65.8	571.3	1.2	9,206.6
2nd Quarter	9,008.0	66.1	0.3	—	9,074.4
3rd Quarter	9,047.7	65.8	—	—	9,113.5
4th Quarter	9,079.8	65.7	—	—	9,145.5
1978					
1st Quarter	8,933.9	74.2	336.1	—	9,344.2
2nd Quarter	7,542.3	69.9	2.1	—	7,614.3
3rd Quarter	7,338.6	71.3	—	—	7,409.9
4th Quarter	7,794.2	71.2	—	—	7,865.4
1979					
1st Quarter	10,018.6	12.6	—	—	10,031.2
2nd Quarter	9,606.0	13.8	—	—	9,619.8

Including value of royalty oil payments in kind and Saudi Arabian Government share in Abu Sa'fah oil field.
Source: SAMA, *Annual Reports*, 1970–1979.

Table 7.4. Government actual revenue and expenditure (million riyals)

Fiscal Year	1974/1975	1975/1976	1976/1977	1977/1978	1978/1979
Total revenue	100,103	103,384	135,957	130,659	132,871
Oil revenue	94,190	93,481	121,191	114,042	115,518
Other revenue	5,913	9,903	14,766	16,617	17,353
Total expenditure	32,038	81,784	128,273	138,027	147,400

Source: SAMA, *Annual Report*, 1979.

revenues, the question is very meaningful indeed. In fact, about 60 per cent of the country's GNP is attributed to the oil industry, and a major portion of the remainder is generated through activities created and funded by the oil revenues.

Sooner or later the cost of mining oil will make squeezing out the last few barrels greater than the price of other sources of energy. And, as a result, the oil revenues are expected to flow only for a limited and uncertain time period. If part of the revenue is not saved and invested, future consumption will fall. The problem that Saudi Arabia faces is, in principle, not different from the one that an individual who is in his prime working age faces. The objective is to maximize the country's total wealth rather than 'current income'.

Saudi Arabia can maximize its total wealth by investment projects that generate future income. But, since domestic productive capacity could not be expanded at the same rate at which the oil revenues grow in certain years, part of the total investment funds must be invested abroad. The proper question that economic policy makers should ask themselves is what proportion of total investment should be allocated to each type.

The answer to this question is provided in Figure 7.1. Investment funds should be allocated in such a way as to equate the marginal rates of return on domestic (r_d) and foreign investment (r_f). Given the total amount of funds for investment GF, maximum return is achieved by allocating GS to domestic investment and SF to foreign investment.

In the previous section we analysed the problems associated with domestic investment. The conclusion was that the relative backwardness of the economy limited the amount of expenditures on development projects that could be undertaken without substantially increasing the inflation rate during the years of rising oil revenues; for example, 1974–1976 and most likely 1980.

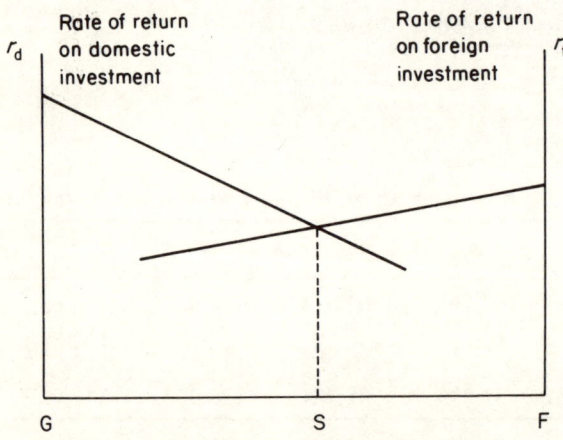

Figure 7.1. Allocation of investment funds.

But is it not more sensible, one may wonder, for the government to shift funds immediately from domestic to foreign investment once the real rate of return on domestic investment falls below the rate of return that could be generated on foreign investment? Not really. Once one takes into consideration some problems that are associated with foreign investment, then almost any positive rate of return on domestic investment becomes attractive.

Here are some possible problems: (1) American and Western European governments may levy taxes on foreign investments—including Saudi investments; (2) the threat of expropriation to coerce the Saudi government to take a particular political stand (the probability of such action may not seem very high at the present time, but nevertheless positive); (3) the more likely course of action to exert political pressure is simply to block the movement of Saudi funds.

One thing is certain: Saudi Arabia's investment in Europe and America reduces its political freedom and increases the political leverage of the foreign countries where its assets are invested. In other words, property rights of the Saudis in their investments abroad could be threatened at any time.

Under normal circumstances, however, the greatest threat to Saudi foreign investment is the world rate of inflation. No investment is immune to inflation. But the threat of inflation to Saudi investment in the form of government and private bonds—near money assets—is greater than its threat to investment in real capital goods such as plants and equipment.

In June of 1977, the Chase Manhattan Bank had estimated at 7 per cent the average of the inflation rate in the countries where the members of OPEC made their foreign investments. It had also estimated the annual rate of return on OPEC foreign investment at around 8 per cent. Therefore, the real rate of return on OPEC foreign investment may have been 1 per cent! (The rates of inflation and nominal rates of return are much higher at the present but the *real* rate of return is still about one per cent.)

One of course may argue that a member of OPEC, say Saudi Arabia, could get a greater return on its investment if it did not confine it to near money assets. That is true but, when the Saudis make an investment, say (R), in period (t), they may not get, in period $(t + 1)$, $[R_{t+1} = R_t(1 + r)]$ because there is some risk, however small it may be, that they will get nothing in the next period. What they will get next year is not one hundred per cent certain.

Thus, if we say there is a probability $q < 1$ that no one will threaten the Saudi investment, then, in period $(t + 1)$, the Saudis expect to get

$$E(R_{t+1}) = qR_{t+1} + (1 - q)O$$

$$\hat{R}_{t+1} = qR_{t+1}$$

Therefore

$$R_{t+1} = R_t(1 + r)$$

only

$$\hat{R}_{t+1} = qR_{t+1} = R_t(1 + r)$$

or

$$R_{t+1} = \frac{R_t(1 + r)}{q}$$

In other words, the Saudis keep their investment in the form of near money assets that generate a low rate of return because they are not one hundred per cent certain that their property rights will be protected. Under such circumstances, Figure 7.2 describes the Saudi allocation of investment funds between foreign and domestic markets.

The purpose of using these mathematical symbols was to formally show that as long as there is a chance that somehow the Saudi investment abroad might be threatened, the Saudis should not increase their foreign investment simply because the rate of return on investing in foreign countries is higher than the rate of return that would be generated by domestic investment.

One would conclude, then, that if the Saudi objective is just wealth maximization they should produce no oil in excess of the amount needed to finance domestic expenditure if they expect the net value of their oil to appreciate at a rate that is greater than the rate they are getting now on their foreign investment, which is about one per cent in real terms.

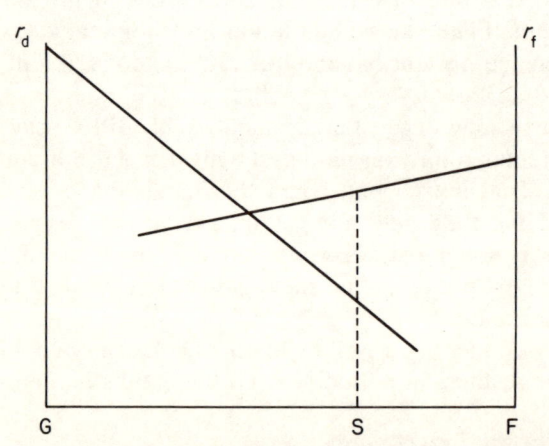

Figure 7.2. The allocation of investment funds under conditions of risk to property rights (r_f and r_d are riskless rates of return).

The actual behavior of Saudi Arabia, however, shows that wealth maximization is neither the only nor the most important consideration for deciding rates of oil output.

The Saudis are worried about the increased influence of radical movements both within and outside the Middle East. And they seem to think that the economic damage which occurs as a result of rising oil prices only serves the political interest of those who seek to destroy the world's stability.

If Saudi Arabia's objectives were only wealth maximization it would not have increased its oil output during the fall of 1978 to avert price rises. The average daily rate of output was 7.2 million b/d in August. When political upheavals in Iran resulted in disruptions in supply, Saudi output was increased to 10.3 million b/d during the months of November and December of 1978. The magnitude of output increase (43 per cent) was staggering and obviously was designed to eliminate the effect of the reduced Iranian oil supplies on the price of oil.

7.4 OPEC AND SAUDI ARABIA

No matter how persuasively one may argue, and no matter how often the argument is repeated, that OPEC is not a cartel, most people, including the majority of professional economists, are convinced that OPEC in fact is a cartel.

In this book an attempt was made to show both the professional readers and the general public that OPEC may only *seem* to be a cartel whereas in *reality* it is *not*.

As a result, the author believes that the *best* way to convince the world that OPEC is not a cartel is to *end* its existence.

There was a time when the OPEC organization served the interest of the oil-exporting countries very well; and that was in the 1970–1973 period, by allowing them collectively and individually to transfer crude oil ownership from the foreign companies to the host countries. That was OPEC's greatest accomplishment.

Today, OPEC pretends that it sets the world oil price. But even when its members do agree on the 'posted' price, and many times they did not agree, they individually raise or lower this price according to market forces without paying too much attention to the agreed-upon prices. OPEC simply has outlived its economic usefulness.

What in fact determines the world oil price is the world's total supply of oil and total demand for it. And if OPEC affects the price, it must either influence demand or supply. Obviously it has no influence on total demand. OPEC as an organization does not have any influence on the individual countries' rate of oil output either.

Of course the total oil output of the members of OPEC does affect the world's price of oil, but this output is determined by the individual countries' behavior rather than by any orders from OPEC as an *organization*.

Because the world has a more or less false image of the oil market, dismantling OPEC will serve the interest of producers (particularly Saudi Arabia) as well as the interest of consumers (particularly the US).

If OPEC disappeared, policy makers in the US, for example, would realize that oil prices are rising simply because demand is growing at a faster rate than supply and *not* because of OPEC efforts to reduce supply and thus raise prices. And once policy makers everywhere became aware that prices are rising due to market forces, then the transition from using oil as an energy source to using other sources of energy would be easier and less disruptive.

Saudi Arabia, on the other hand, would benefit a great deal from dismantling OPEC! The reason is simple: because of international pressure Saudi Arabia has been producing more oil than its economic well-being requires. In addition, and almost always, Saudi Arabia's oil price is lower than the price of anyone else's. In short, Saudi Arabia has been overproducing and underpricing.

Saudi Arabia and every other member of OPEC should sell their oil, just like other non-OPEC oil producers, to the highest bidder. Why should oil be different from wheat, copper, cotton, or any other commodity?

No one would blame US farmers if the price of wheat rises and no one ought to blame Saudi Arabia, *if the alleged OPEC cartel did not exist*, and the price of oil rose.

Of course, selling Saudi oil to the highest bidder will not prevent Saudi Arabia from adjusting its oil output if it believes that doing so will serve its own interest or that of the world at large. But the behavior of the world's oil price during the entire year of 1979 and the winter of 1980 showed that there was very little that the Saudis could do to influence significantly the oil market.

A *WORD* ON THE APPENDICES

The Appendices to this chapter are devoted to an elementary discussion of some of the topics that have been mentioned in varying degrees of detail.

'The Nature of Inflation in Saudi Arabia' and 'The Price of Land in Saudi Arabia' were published in *Saudi Business*, a weekly periodical that is published in Jeddah, Saudi Arabia. 'Private Saudi Investment Abroad' was published in the *Arab News*, a Saudi daily newspaper. 'Not For Rent Please' is based on an article that I wrote in Arabic and was published in *Al-Riyadh* newspaper.

APPENDIX A
THE NATURE OF INFLATION IN SAUDI ARABIA

Perhaps there is not a single economic topic which receives more attention but is awarded less understanding than inflation. The reason inflation claims the headlines is rather obvious—it affects in varying degrees all of us. And it is not properly understood because, like many other economic problems, it is not what it appears to be.

Inflation is a *continuous* rise in the *general* price level. Not every price rise is inflationary. It is the normal consequence of resource allocation in a market economy that some prices rise and others fall. Only when the overall level of prices continuously rises do we have inflation.

But a sustained overall rise in prices could not exist unless there were a sustained rise in demand to finance the additional expenditure. And the aggregate community demand could not continue to rise unless there were a corresponding increase in the means of payments of a community. The most relevant means of payments are currency in circulation and demand deposits.

A continuous increase in the means of payments does not induce inflation if there is a corresponding increase in the supply of goods and services. That is to say, if the increase in purchasing means is matched by an increase in the availability of what is to be purchased, there would be no inflation.

Therefore, in Saudi Arabia, or for that matter anywhere else, inflation occurred—and can continue to occur—because the supply of money (the source of demand for goods) increased at a faster rate than warranted by the actual increase in the supply of goods and services.

In most countries there is a distinct difference, though there is a relation, between money creation and the government budget. In the Saudi case, however, the source of the means of payments is the public sector. How? The government finances its expenditures by writing checks on SAMA. SAMA converts the checks into Saudi Riyals, and then the Riyals enter private circulation.

For the past eighteen months or so (the entire year of 1978 and the first half of 1979), Saudi Arabia has been experiencing a very low rate of inflation compared with its rates between 1974 and 1976. According to SAMA, inflation had reached about 40 per cent in 1974/75, and it is now (July 1979) about 10 per cent.

I think that both figures are inaccurate, because of the statistical method which is followed to get them. I have a strong suspicion, which is based on technical and intuitive reasons, that the inflation rate in 1974/75 was much more than 40 per cent and is much less than 10 per cent now.

Regardless of the degree of accuracy of these figures, one thing is very clear: the present rate of inflation is only a fraction of the 1974/75 rate. What happened? To answer this question, we must try to understand why the supply of money in Saudi Arabia was allowed to rise by so much so as to cause that high level of inflation.

As a result of the January 1974 rise in the price of oil, the government oil revenues had increased by an unprecedented rate. And since oil is a non-renewable resource, government expenditure on development projects, in an effort to facilitate the creation of other sources of income, was greatly increased. But the increase in government expenditure entails an increase in the means of payments, and that results in an increase in total demand.

Of course, the total Saudi demand for any internationally traded commodity is such a small fraction of its total supply that it could not affect its price. For example, if the Saudi demand for GM cars was doubled, it would still be a very small fraction of GM total output.

Then, given that most of the goods that are demanded by the Saudi economy are imported, and given that the total quantity of imports is a very small part of the total world supply, why were high rates of inflation experienced?

The brief answer is this: imported goods could not be dropped from the sky. They have to come through seaports, airports, and on highways. Unfortunately, the supply of these things, the 'infrastructure', could not be increased at the same rate at which the demand for imported goods was rising.

The country's highways, seaports, and airports got congested. There was a time when ships had to wait as long as six months before they could be unloaded. Things were not much better in the highways or in the airports. The supply bottle was full but its neck was not large enough to allow the passage of the required demand.

Finally the high rate of inflation fell to its present rate (June 1979), which is not above the general world rate, because of two main reasons. First, the government budget—the indirect source of inflation—stayed constant in nominal terms and fell in real terms. Second, the substantial improvement of the Kingdom ports increased the width of the neck of the bottle and allowed more goods to come in.

There is a widely held belief that inflation is caused by a rise in cost. This is a myth that nevertheless has a strong influence on economic policy makers throughout the world. This 'cost-push' theory of inflation is popular because it is based on another popular but false theory, namely that costs determine prices.

If you ask a businessman, who is a very successful one, 'how do you determine your price', most likely he will tell you, 'by cost plus a profit margin'. But he is wrong. It is possible to be very good at something without knowing how it works. Some of the world's best car drivers may hold false theories about the internal combustion engines.

If a businessman claims that he determines his prices by cost plus a certain margin for profits, then what determines the size of the margin. Let us say he tells us that his goods prices reflect their costs plus 25 per cent profit margin, then, we will ask, why 25 per cent? Why not 1000 per cent or 5 per cent? Or anything else?

Obviously, there is something else that determines prices in addition to costs, and that is the market demand and the action of competitors. Costs are important only in one direction—downward. No business firm can continue to exist for long unless it recovers at least its average costs. But once the price is above costs, then the level of demand and the extent of competition are the two forces that decide market prices.

Let us use the following hypothetical situation to explain why 'cost-push' is so appealing as an explanation for inflation: Assume that the demand for the meat of locally raised sheep rises for some reason. All that your local butcher knows in the first week after the rise in demand is that he runs out of his local meat variety at an earlier hour of the day than usual. Once he realizes he can sell more meat, without reducing his prices, he tries to increase his purchases of meat. But every other butcher is doing the same thing. The man who brings the sheep to the market also sells more. He can't, however, bring more sheep unless he pays more to the sheep raisers. The price of sheep rises, then the price of the meat follows.

You complain to your local butcher about the increase in his prices. He tells you quite correctly that his costs rose. The butcher's costs did rise, but what caused that was the original increase in demand. (The above example was borrowed from Alchian and Allen, 1974: 95–96.)

Clearly costs may rise for reasons other than an increase in demand. But that is not inflation. Inflation is not a one-shot increase in price; it must be remembered that it refers to a *continuous* rise in the overall level of prices. And if costs rise continuously, sellers will not be able to pass on their costs to consumers unless the consumers' money income is continuously increasing to finance the additional expenditure.

In short, demand must be continuously rising at a faster rate than supply in order for inflation to be sustained. But demand cannot persist in exceeding supply, unless it is a nominal demand. And that can occur only if the economy's total liquidity is rising faster than its capacity for providing goods and services.

APPENDIX B
THE PRICE OF LAND IN SAUDI ARABIA

It is widely believed that the 'outrageous' land prices, in a country with a large area and small population, defies the laws of supply and demand. But nothing could be further from the truth. The price of land, like the price of anything else, is determined by the market forces of supply and demand. And if it appears otherwise, that is because such concepts are rarely properly understood.

People demand land for direct use—as a site for a buil⋯, or a factory. But they also demand land as a hedge against inflation or even as an investment when they expect the price of land to rise at faster rate than other 'real' goods. Thus, the rapid urbanization of the country, the recent high inflation rates, and the huge increase in disposable income all combined to increase the demand for land.

Yet the supply of land, contrary to the popular myth, is very limited, since land will be valuable only if it has other things that make it useful such as water, electricity, and proximity to metropolitan centers. And if the price of land in locations with no utilities rises, that means people expect utilities to become available in the future.

Because it takes money, time, and effort to increase the supply of water and electricity and to build roads that connect different parts of a city, the supply of useful land in Saudi Arabia is scarce. In fact it is very scarce, since land, unlike many other goods, cannot be imported.

Further, the absence of a capital market in Saudi Arabia, made land the most attractive form of investment into which many people channelled their savings. Inflation has taught many investors that putting their money in saving accounts is not an attractive alternative to owning land.

Recently (1979) the prices of land stopped rising, or only rose at very slow rates, precisely because demand has slowed. And demand for land decreased for two reasons. The first, and most important, is that the supply of money, which finances the country's total demand, has been increasing at a much smaller rate than before. The second reason is the gradual realization that other 'productive' investment such as manufacturing could also result in high rates of return.

In short, the prices of land are high simply because the demand for land was and is large and because the supply of desirable land, relative to the demand, is very small.

If it is desirable to further reduce the price of land, then what could be done to accomplish that?

The obvious answer is to increase its supply. Not physically, of course, since there is a lot of land to be had at almost any price if one wants land 100 kilometers from the urban centers.

The supply of useful land could be expanded by generating more electricity and pumping more fresh water. The problem with that is this: neither electricity nor water is free. Their supply could be only increased at the cost of giving up something else.

It is that old evil, which economists call scarcity—and which means we can increase the supply of a useful good only if we are willing to forego something else—that prevents us from increasing the supply of useful land.

Land is, therefore, and even in Saudi Arabia, a very scarce commodity. Those who insist that desirable land in Saudi Arabia is not scarce, or at least ought not to be, must be thinking of the apparent physical abundance of land in this Kingdom. But thinking of land in terms of the total area will lead one to false conclusions. The land which commands a high price is an extremely small portion of the total area. And that is the portion which is very scarce indeed.

APPENDIX C
PRIVATE SAUDI INVESTMENT ABROAD

The casual observer of the Saudi economy must be wondering why a large number of private Saudi citizens are investing some of their savings abroad, where the average rates of return on their investments are far less than what they can realize within the Kingdom. To be sure, the bulk of private savings is still being employed to produce domestic goods and services. Nevertheless, any amount of domestic funds that seeks foreign outlets, at the time when the Saudi government is doing its utmost to encourage private domestic investment, must be explained.

It may take years, if not forever, before we know *the full* explanation. But here are some of the factors that may explain this behavior.

First, individuals and business firms may invest abroad for the sake of diversification. That is, an investor may not want to put 'all of his eggs in one basket'. Thus he may prefer a lower average rate of return on his investment abroad and on his investment at home to a higher rate of return on domestic investment alone.

The majority of investors are not worried about political instability or risks of expropriation in Saudi Arabia. By any standards, Saudi Arabia is politically a very stable community, and its commitment to the respect of private property rights is firm and certain. What may worry investors, however, are economic forces that are beyond Saudi Arabia's control, like fluctuations in oil prices and political upheaval in other countries.

Second, financial rates of return on domestic investment in Saudi Arabia are certainly higher than what they are in Europe or in America, but that does not tell the whole story. A businessman, like everyone else for that matter, does not only think of his rates of return but is also concerned with the amount of the non-financial costs he has to incur before he reaps the financial fruits of his efforts.

And in Saudi Arabia, non-financial costs are substantial. Among them is that a Saudi businessman has to work longer hours and under less favorable conditions than his counterparts in Europe and America. The reason: he has to do almost everything himself. Able assistants are either very expensive or simply not available at any price.

There are also other problems that Saudi businessmen must contend with. Not the least of which is the familiar, but nevertheless very annoying, bureaucratic red tape. For example, a contractor may spend hours of his valuable time waiting to see a petty clerk who will not immediately see him and who is causing the delay of long overdue payments.

We can also think of the difficulties that a businessman faces before he is able to get entry visas for his expatriate employees. That is not to say that entry visas should be granted immediately and easily since there may exist good security reasons for the delay. It is to say, however, that the long process which is required before visas are granted does increase the cost of doing business in Saudi Arabia.

The high administrative costs of doing business in Saudi Arabia may explain why many foreign companies establish their regional headquarters outside the Kingdom even when it is the biggest market for their products.

In short, some Saudi investors are willing to forego some financial return for convenience and ease of doing business in other countries.

Some commentators have alleged that academic economics is irrelevant to a businessman because it is only concerned with the financial side of doing business. But that is not true. When an economist tells you that a businessman, just like any other member of society, maximizes his utility rather than his wealth, what he precisely means is that people worry about other things in life in addition to wealth.

Third, sometimes foreign investment is undertaken to increase a domestic firm's foreign prestige and thereby its future domestic income. That is, a firm may be interested in a foreign venture for the sake of establishing connections with foreign firms. Very frequently, the help of more experienced foreign firms is badly needed by domestic firms for the purpose of ranking and successfully carrying out their domestic investment projects.

Fourth, traditionally, land was the most desirable outlet for private excess funds in Saudi Arabia. Many people have learned that investing savings by buying land takes very little effort, requires no exceptional talent, and resulted in high rates of return. But toward the beginning of last year (1978), the value of land had ceased to increase, and that eliminated buying land as a means of investment.

To recapitulate, Saudi investors will find it desirable to invest some of their cash abroad even when the rates of return on similar domestic investments are higher. They do so because the purpose of investing is to maximize one's wealth with the least possible 'headache'. And investing in domestic Saudi projects, in spite of the government's effort to make it worth while, may result in high financial rates of return, but will also require greater personal efforts on the part of the investor.

Some of those who chose to invest outside the Kingdom must have decided that the additional non-financial costs of doing business here are high enough to compensate for the lower rates of return on their foreign investment. If that is not the motive of some other investors, then they may have decided on foreign investment for diversification purposes or for the purpose of acquiring foreign know-how.

APPENDIX D
NOT FOR RENT PLEASE: THE CONSEQUENCES OF RENT CONTROLS

The observer of the real estate business in Saudi Arabia will not fail to note that rents of residential properties are (in 1979) gradually falling. But he will also find that many property owners have refrained from renting their newly constructed houses and apartment buildings.

In the last few weeks (Spring 1979) the Saudi newspapers have printed many stories about the increasingly large number of unoccupied residential units. According to one estimate, in Jeddah alone there are about fifty thousand—the figure was attributed to the Jeddah Mayor, Mr. Farsi—units of empty housing.

What is going on? Why are the owners of these buildings and houses foregoing an amount of rent which is very large by the rest of the world's standards? Some commentators have even claimed that property owners in Saudi Arabia are just plain irrational!

Economic analysis, however, teaches us that this phenomenon of unoccupied housing is natural and to be expected. Why and how?

There are at least two important reasons for this state of affairs.

First, and foremost, there is the existence of rent controls in Saudi Arabia. Of course, as is well known, newly constructed housing could be rented at the highest rent that the market forces allow. But what makes owners of houses and apartments hesitate before renting is their fear that in case rents rise in the future, they cannot revise the previous contracts and demand higher rents. Thus, they decided to forego some rental payments until they are sure that lower rents are here to stay and are not a transient phenomenon.

Secondly, those who built the newly finished houses would not have built them if they had not expected a certain level of rents. But once they put up these houses in the rental market they could not get the rental amounts which they expected. Rents in general fell because (i) everyone expected higher rents and thus started construction, and that increased the *supply* of housing; (ii) the growth of general *demand* for everything—including housing services—had slowed down a great deal because of the reduction in government spending.

In fact, one of the most important factors for the large increase in the supply of housing was the government help and encouragement through the Real Estate Development Fund.

However, what is relevant here is that because the housing supply has grown faster than housing demand, rents fell or at least failed to rise.

But all that an owner of a property knows is that he is not getting the rents he was expecting. He does not know if lower rents are permanent or not. Thus he waits in order to be sure that he is not making a mistake by renting immediately.

It is a fact well known to students of economics that when prices—including prices of housing services (rents)—are rising, sellers immediately accept them; and during periods of falling prices, sellers hesitate to sell and postpone their decisions. Why?

As prices rise, every individual seller thinks he is the only one who is obtaining the higher prices, thus he sells right away thinking he is making more money than others.

As prices fall, on the other hand, each individual seller knows he is getting less than what he was expecting. So he thinks he is the only one who is getting less. His natural reaction is to wait in order to make sure that he can't be made better off by waiting any further.

In other words, information is not free. And sellers pay for it by foregoing sales and rental revenues until they determine the true state of the world, and then they make their decisions accordingly.

Therefore, many people in Saudi Arabia have decided to keep their houses and apartments 'locked' and unoccupied because of the influence of rent controls in case rents rise in the future and because of the steady decrease in rents.

Rents are obviously still very high, but not as high as they were in 1974–75. As a result of the decrease in rents coupled with the existing rent controls, many housing units are not available at current rents.

Owners of residential housing will finally learn that the 1974–75 levels of rent are gone forever and will accept lower rents if rent controls are not slowing this process.

The most effective way to reduce the number of unoccupied newly constructed villas and apartments is to permanently revoke rent controls.

REFERENCES

Alchian, A. A. and Allen, W. R. (1974). *University Economics*, Prentice-Hall International, London.
Adler, J. H. (1965). *Absorptive Capacity*. The Brooking Institution, Washington, D.C.
Chase Manhattan Bank monthly report (June 1977). *Business in Brief*, New York, N.Y.
Knauerhase, R. (1975). *The Saudi Arabian Economy*, Praeger, New York.
SAMA, *Annual Report* (1970–1979). Riyadh, Saudi Arabia.

CHAPTER 8
Saudi Institutional Processes and Economic Modernization

INTRODUCTION

The Kingdom of Saudi Arabia was born on September 22, 1932, when King Abd-al Aziz Ibn Saud was able finally to unite more than three-quarters of the Arabian Peninsula under his rule. He started his effort on January 15, 1902, when he captured Riyadh—the present capital of the Kingdom—with the help of about fifty armed men.

King Abd-al Aziz's accomplishment was a tall order when one realizes that he transferred a huge country from essentially a group of constantly feuding tribes and towns into an integrated nation state.

By the end of the seventeenth century the great Moslem empire had completely disintegrated. And in the Arabian Peninsula, the original home of Islam and its great Prophet Muhammad, Islam itself had almost disappeared.

King Abd-al Aziz was able to revive Islamic zeal among the population, particularly amongst the nomadic segment of it which in the words of Ibn Khaldon, the fourteenth-century Arab scholar, 'are difficult to lead except when they follow a prophet or holy man for religion alone can diminish their haughtiness and restrain their jealousy and competition' (Cole, 1975: 25).

Before King Abd-al Aziz no one was able to gain the respect, loyalty, and, more significantly, the obedience of the Bedouin tribes. Thus when he gained their trust mainly through religious persuasion, he also became the leader of the most powerful army in Arabia because in Bedouin culture only courage and hospitality are important. When King Abd-al Aziz needed the Bedouins as soldiers he knew, therefore, that they were trained to do nothing as well as fighting. In short, the Saudi King combined the Koran and the sword, and that was the secret of his enormous success.

When King Abd-al Aziz Ibn Saud died in 1953 his successors understood perfectly well that religion had played a crucial role in Saudi Arabia, not only because the Saudi cities of Mecca and Medina are the holiest cities in the Islamic world, but also because religion was the most important pillar on which the Saudi Kingdom was built since the earliest days of its beginnings.

Today, religion is the most important factor in Saudi culture, and the large number of people who have been unable to understand why it influences everything in Saudi life the way it does, just failed to appreciate that Saudi Arabia owes its national identity to Islam.

8.1 DECISION MAKING

In Saudi Arabia the King is the head of state, prime minister, commander-in-chief of the armed forces and the chief executive of the government. In the final analysis he is the ultimate authority.

It has been documented, nevertheless, that Saudi leadership since the early days of the creation of the Kingdom of Saudi Arabia has tried its best to reach a consensus among the different groups of which the population is composed.

The number of the members of the Royal family, which is perhaps around five thousand, is larger than the number of all the members of many political parties in many of the less developed countries. And it seems certain that the senior members of the Royal family (Al Saud) as a group have a greater influence on the King than any other group of Saudi citizens. It is also true that all the members of the Royal family are accessible to the average citizen and therefore are in a position to know the popular sentiment toward different issues.

Collectively the members of the Council of Ministers, which functions as the instrument of royal authority in legislative as well as executive matters, also have great influence in formulating and/or executing Saudi policy.

Since 1976, the Council of Ministers has been composed of a first deputy prime minister (Crown Prince Fahd), a second deputy prime minister (the head of the Royal Saudi National Guard, Prince Abd Allah), and twenty other ministers.

The ministers of defense and interior are usually chosen from the most prominent members of the Royal family.

The majority of the members of the Council of Ministers are not, however, members of the Royal family and were chosen mainly because of their individual expertise. Western readers are perhaps more familiar with the names of Ahmad Zaki Yamani (Petroleum), Hisham Nazir (Planning), and Ghazi Al Qusaibi (Industry and Electricity) than the names of many of the Royal family members who serve in the Council of Ministers.

But what many foreigners do not understand is that a Saudi Minister is more of an assistant to the King than a cabinet member who formulates policy. His functions in a way are similar to that of a US cabinet member who is in fact an assistant to the president and is quite different from a cabinet member in most of the European countries.

For example, petroleum policy is really *not* made at the Ministry of Petroleum but rather in the King's office and in the office of the Presidency of the Council of Ministers, where the Supreme Petroleum Council ponders petroleum issues. The function of the Petroleum Ministry is mainly to provide information, if such information is needed, to formulate policy and to execute whatever policy decisions were taken by the country's leadership.

That does not mean that the individual members of the Council of Ministers have no power; many of them do. It only means that, in general, the power of the individual members of the Council of Ministers depends, to a large extent, on his personal ability to influence the political leadership as well as his own colleagues, the other fellow ministers.

The source of power that is derived from the office of a minister in most cases stems from the fact that ministers, as individuals, head large and powerful bureaucracies that are usually able either to implement government policy or circumvent it.

8.2 THE MINISTRY OF FINANCE AND NATIONAL ECONOMY: A GOVERNMENT WITHIN GOVERNMENT

We have mentioned that the power of the majority of the members of the Council of Ministers is not intrinsic to their offices and depends to a large extent on their personal abilities. An outstanding exception to this is the man who heads the Ministry of Finance and National Economy. It does not matter who he happens to be; he has enormous power simply because he is the Minister of Finance and National Economy. Of course, if he also happens to be an effective executive and/or an influential man, he will be even more powerful.

This ministry allocates funds and oversees their expenditures. It does not only prepare the budget, it also must approve the expenditure of funds even after they have been allocated. And in cases when its officials could not stop an already allocated amount, they can delay the date of its payments, which is enough to make every government official do everything he can to please the men who run this incredibly powerful ministry.

It ought to be said, however, that although this ministry is very powerful, it has almost no influence on the appropriation and payments of funds in the area of national security, which receives the highest priority and is not governed in a substantial way by economic considerations.

Outside Saudi Arabia the ministers of Petroleum, Planning, and Industry are far better known than the Minister of Finance and National Economy; yet, he is certainly far more powerful in influencing the daily actions of the government.

The Saudi economy depends on oil. And oil is entirely owned by the government. As a result, the Saudi government has perhaps more power than any other government in the world in controlling the country's economic activities—particularly the flow of income. It is true that the Saudi government is also very rare among governments everywhere in its dedication to private enterprise. But the private sector depends on the oil revenues which could only reach it through governmental channels.

And the government agency that receives and disburses oil revenues is the Ministry of Finance and National Economy.

Theoretically, the Ministry of Planning sets economic goals and formulates yearly economic policy, but its ability to do so is constrained a great deal by the fact that the Ministry of Finance and National Economy not only prepares the entire government budget but also controls monetary policy.

The power of the Ministry of Finance and National Economy over monetary policy stems from its control over all the specialized government credit institutions (e.g. the Industrial Development Funds, the Real Estate Development Fund, etc.)—which are a significant source of liquidity—and from the fact that its minister has a lot of influence (chiefly indirect) over SAMA, the Saudi equivalent of a central bank.

If we compare the Saudi government to its American counterpart, the Saudi Ministry of Finance and National Economy would be equivalent to the following combined American agencies: the Department of the Treasury + the Federal Reserve System + the Office of Management and Budget. And to appreciate the enormous power of the Saudi Ministry of Finance and National Economy, one has to remember the extent and scope of Saudi government influence on all aspects of the economy.

To be the minister of Finance and National Economy in Saudi Arabia is a truly demanding job. Not just because it is an administrative nightmare, but also because it is impossible to please even a small fraction of the economic interest groups who compete to get as much as they can from the government revenues. It is in a sense a thankless job. Yet, it is also true that whoever runs the Ministry of Finance and National Economy is in fact running a government within the Saudi government.

8.3 SAUDI ECONOMIC DEVELOPMENT

Prior to the unification of Saudi Arabia in 1932, an integrated national economy did not even exist. Economic activity outside Hejaz (where the holy cities were located) was confined to raising livestocks by Bedouins, primitive agriculture, and production of simple tools by craftsmen who lived in small towns which were concentrated around sources of water.

In Hejaz (the north western part of the country) an opportunity did exist for an economic development that could have affected all of Saudi Arabia because of the yearly influx of pilgrims visiting the holy places in Mecca and Medina. But the lack of security made movement of people and goods an exceptionally risky activity.

The justified fear of raids on camel caravans by ruthless Bedouins, the infrequency of rain, and the inhospitable weather of the desert limited the size and scope of economic activity and made production feasible only on a small scale for small markets and practical subsistence in nature.

By the end of 1932, the security problem was completely solved. And from that date until today Saudi Arabia has been the safest place on earth for both property and people.

Oil was discovered in commercial quantities in March 1938, but World War II interrupted the development of the petroleum industry. But in the period that immediately followed the end of the war, production increased rapidly. Total output rose by more than three-fold from 60 million barrels during 1946 to 200 million barrels during 1950.

Since the end of World War II oil has been the source of revenues for both the private and public sectors.

The overall performance of the Saudi economy can be accurately measured by the growth of the government budget—the vehicle through which oil revenues reach every segment of the economy—which increased from $103 million in fiscal year 1947/1948 to almost $48,000 million for fiscal year 1979/1980.

But the highest rates of growth were achieved from 1974 to 1980 (see Table 8.1). In this period the main problem was the level of inflation, which was due to the high growth of liquidity, given the economy's modest infrastructure which made it impossible for imports (supply of goods and services) to increase as fast as nominal income.

By the end of 1979, however, the battle against inflation was won and its current rate (1980) in Saudi Arabia is below the world average. The reason: the drastic decrease in the level of liquidity (Table 8.2) and the improvement of the country's seaports which resulted in decreasing the waiting periods of incoming loaded ships from six months in 1975/1976 to less than six hours in 1978/1979.

It is planned that by the end of 1980 a total of 114 piers be in operation at the five main ports, as compared with only 57 in 1977.

Table 8.3 gives a bird's eye view of the economic development of the country during the 1970s. The projects that received the government's

Table 8.1. Real GDP (annual growth rates)

	1975/1976	1976/1977	1977/1978	1978/1979
GDP	8.6	14.8	5.9	7.6
Oil Sector	1.1	13.2	−0.5	1.8
Non-oil Sector:	19.8	16.9	13.8	13.7
Private	17.8	18.9	13.9	14.1
Public	23.9	12.9	13.5	12.9

Source: SAMA, *Annual Report*, 1979.

Table 8.2. Annual growth rates of money supply and real supplies of goods and services

Fiscal year	Money supply (M_3)	Real supplies of goods and services
1975/76	73.9	34.5
1976/77	52.7	30.3
1977/78	43.6	23.9
1978/1979	14.5	15.7

Source: SAMA, *Annual Report*, 1979.

greatest attention are road construction and port, airport, and telecommunication schemes, housing, the establishment of new industries, and the building of schools.

The improvement and expansion of roads is very impressive by any standards, but it is even more so when one remembers its state only a few years ago. In 1968 the total cumulative length of Saudi Arabia's asphalted roads was less than 2000 kilometers; by the end of 1979 the total had risen to more than 20,000 kilometers (see Table 8.4).

The accomplishment in the field of education is just as impressive. The number of students that were enrolled on all levels of educational institutions in 1979 had reached 1.35 million, as compared with only 42 thousand in 1952.

In the period 1975–1977 there was an acute housing shortage which was made even worse by rent controls. For example, the monthly rent of a two-bedroom unfurnished apartment in 1974 in the major cities was, on the average, between $150 and $200. The same apartment in 1977 would have been rented for somewhere between $500 and $800 per month. During 1980, that apartment was rented for less than $300.

The housing shortage has been eliminated chiefly because of the establishment in 1976 of the Real Estate Development Fund which grants interest-free loans to citizens for building their own dwelling as well as for constructing residential buildings for others to occupy. By the end of 1979 the Fund's total loan disbursements since its creation in 1976 reached Rls 24.4 billion.

Telephone quality and service has been improved rapidly during the 1970s. By the end of 1981 work is expected to be finished on the 470,000 automatic telephone lines operating on electronic exchange in 72 towns throughout the Kingdom. In addition, a total of 291,300 lines were already in operation at the end of 1979.

Telex lines are expected to reach 9,000 by the end of 1980, as compared to 6,350 lines in mid-1979. This service is now available in 19 cities.

There is no other less developed country which did what Saudi Arabia is doing to further the industrialization of the country by the private sector. And given the trader's mentality of the majority of Saudi businessmen,

Table 8.3. Project budget expenditure (million riyals)

	Fiscal year					
	1974/1975	1975/1976	1976/1977	1977/1978	1978/1979	1979/1980
Total project expenditure	26,397.0	74,379.0	74,433.4	74,866.0	83,047.7	105,680.0
Council of ministers and related budget headings	1,658.9	4,761.8	4,756.9	4,924.5	4,399.4	13,964.0
Municipal and rural affairs	3,683.8	13,221.6	14,758.0	11,681.3	7,966.8	9,798.8
Public works and housing	114.4	185.7	9,061.4	7,856.8	5,649.4	3,022.5
Information	205.3	636.7	959.8	1,064.0	723.5	634.3
Civil aviation	1,150.8	4,469.9	4,469.9	4,370.0	3,912.8	6,804.6
Interior	973.2	2,301.3	3,078.9	3,293.4	3,330.5	4,131.9
Labour and social affairs	165.7	1,408.5	2,040.8	2,237.0	1,452.3	2,126.5
Health	435.1	2,061.6	1,737.0	1,758.3	1,855.0	1,822.0
Education	1,265.6	6,355.1	6,367.6	7,955.3	5,123.1	5,771.5
Communications	4,212.0	10,994.2	15,380.7	7,822.5	7,377.0	9,811.3
Finance and national economy	1,955.3	7,030.1	3,984.8	3,754.3	3,309.5	7,868.3
Industry, electricity, and commerce	114.4	586.7	1,081.0	488.0	377.3	3,450.5[a]
Agriculture and water resources	1,053.5	1,718.0	1,721.4	1,511.4	1,854.4	3,112.0
Public investment fund	3,000.0	1,600.0	—	—	4,000.0	4,250.0
Other	6,409.0	17,047.8	25,396.3	39,002.7	50,433.0	49,379.4
Less: Expected shortfall	—	—	−20,361.1	−22,853.5	−18,676.3	−20,258.6

[a]Including gathering and liquefaction of gas.
Source: SAMA, *Annual Report*, 1979.

Table 8.4. Road network in the Kingdom (cumulative length in km)

Year	Asphalted roads				Rural roads
	Main	Secondary	Feeder	Total	
1974/75	6,141	5,556	473	12,170	8,510
1975/76	7,182	5,798	1,125	14,105	11,193
1976/77	8,362	5,897	1,779	16,038	13,307
1977/78	9,618	5,959	2,660	18,238	16,948
1978/79	10,834	6,030	3,270	20,134	20,119

Source: SAMA, *Annual Report*, 1979.

government encouragement is necessary. The government incentives to the private sector include: 'exemption from custom duties of imported machinery and equipment, spare parts, raw, or semi-processed materials, and packing materials, protective tariffs or quotas against competing imports, financial assistance on a highly concessionary basis, long term leasing of industrial sites at nominal rent, preferential treatment for locally manufactured goods in government procurement and assistance in the identification of viable projects through market research and feasibility studies' (SAMA, 1979: 73).

Yet, it seems that most of the industrial projects that have been undertaken by Saudi industrialists are confined to the assembly of already manufactured parts.

The main problem that hinders private industrial development is the lack of skilled and semi-skilled Saudi workers. It is not unusual when you go to an industrial plant in Saudi Arabia that you will not find a single Arabic-speaking individual, for the managers are usually Europeans and the workers are almost certain to be from the Indian subcontinent or from the Far East.

Nevertheless, the mere existence of privately owned factories of any kind in a country which ten years ago had almost none is very encouraging. From 1975 to 1978, a total of 1,035 industrial establishments were licensed, with a total capital of Rls 16,780 million.

The most massive industrial projects are, however, undertaken by the government under the auspices of the General Petroleum and Mineral Organization (PETROMIN) and the Saudi Arabian Basic Industries Corporation (SABIC).

The biggest project that PETROMIN is engaged in is the construction of 1,270 kilometers East–West crude oil pipeline linking the Ghawar (the world's largest) field to the Red Sea port of Yanbu. Eventually 3.5 million b/d will be the capacity of this pipeline.

The eleven pump stations that were built to push the crude through the pipeline to Yanbu will be fuelled by NGL, which is also supposed to reach Yanbu through 1,168 kilometers NGL pipeline. The NGL pipeline is being built by Techint Arabia Limited at a cost of $104 million and supervised by AAMCO.

The mammoth gas gathering and processing system, which is expected to collect and process about 3.3 billion cubic feet per day (cfd), is being implemented for the government by Aramco. This enormous project was undertaken to utilize the 'wet' gas which is produced in association with oil and which was hitherto flared; roughly 500 million cfd of associated gas is produced with every 1 million b/d of crude oil.

The original cost estimate of this project was $4.5 billion. PETROMIN says it will not cost more than $12 billion, and other oil industry insiders say the figure may reach $20 billion before completion (*Saudi Business*, October 5, 1979: 16).

About 40 per cent of the collected gas will be used to produce fuel gas for industry, power electric generation plants and desalination of sea water, as well as to provide ethane for use as feedstock for the petrochemical industry at Jubail and Yanbu. The other 60 per cent of the gas will be exported in the form of propane, butane, and natural gasoline.

SABIC, in partnership with such giants as Exxon, Shell, Mobil, Dow, Texas Eastern, and a large consortium of Japanese companies, is building huge petrochemical, fertilizer, and metallurgical projects at Jubail and Yanbu, which when completed should give Saudi Arabia a large industrial base and should make it one of the world's largest exporters of petrochemicals, in addition to being the largest crude oil exporter.

The pollution of the air, land, and sea which will undoubtedly accompany these industrial developments is but one of the problems that economic progress brings with it. Saudi Arabia, like many other countries that underwent rapid economic change before it, has to worry about these problems, and has to do so very soon.

8.4 SAUDI ECONOMIC ISSUES IN THE 1980s

By any standards, the growth rates and structural changes that the Saudi economy had experienced in the 1970s, are very impressive. In the previous section those favorable developments were discussed.

In this section we will give a glimpse of the economic issues that Saudi economic policy makers have to contend with in the 1980s and beyond.

(A) Income distribution

No detailed figures are available concerning the distribution of income in Saudi Arabia, yet certain generalizations can be made.

Firstly, the urban population was (and still is) the segment of the population that benefited the most from economic growth, for they are the ones who have the required skills to participate most effectively in the country's development. From this group comes the leadership of business and industry as well as the high ranking government officials. In addition, the urban population benefits the most from the services that the government offers free of charge like education, health services, and the use of roads. And they are the ones who consume the greatest amount of water and electricity, which are available to the city dwellers at a very small fraction of their actual cost.

Secondly, among the urban population income is concentrated in the hands of those who work in the private sector, where salaries and wages are a multiple of what the government pays to its employees.

It is true that the source of wealth in Saudi Arabia is oil, and oil is one hundred per cent owned by the government; but the way this wealth is spread to the population favors those who own business establishments and to a lesser extent those who work for them. And the reason for that is traceable to the fact that those who work for the government receive fixed wages and salaries, the purchasing power of which is continuously falling due to inflation. Those who are in the private sector, on the other hand, sell their services in the open market and are often able to benefit from shortages that are caused by the continuous growth of liquidity, which has often been growing at a faster rate than the growth in the supply of goods and services.

The absence of income taxes and other kinds of taxes only compounded the problem. The only tax that Saudis are asked to pay, and in fact rarely pay, is the 'Zakat', which is 2.5 per cent of an individual's net worth, and that is obviously very small and would have very little effect as an instrument of income redistribution even if it were paid by everyone.

The press in Saudi Arabia and abroad had often talked about many Saudi 'billionaires'. One could never be sure of the accuracy of newspaper stories, but one thing is certain: the number of Saudis who are 'millionaires' is growing, and those who joined this exclusive club did so mainly because they happened to be engaged in some sort of private business activity when the economy boomed in the period 1974–1977.

During the mid-seventies the easiest way, though not the only way, for one to become a millionaire was simply to engage in the selling and buying of land. It was not unusual for the value of a plot of land which an individual bought one day to more than double a week later (see Appendix B to Chapter 7). Yet, there were many, though certainly not all, government employees who were busy doing their job and did not participate in the real estate bonanza.

One can conclude, therefore, that the group of Saudi citizens who benefited the least from economic growth are the Saudi Bedouins. That is not to say that the Saudi Bedouins are not better off today, because they are, it is only to say that their relative share of the pie is small and continuously decreasing.

But how many Bedouins are in Saudi Arabia? Estimates have varied from as high as 90 per cent to as low as 10 per cent (Said, 1979: 137).

The reasons for these different estimates are many, but the chief one seems to be the confusion between Saudi citizens who are descendants of tribal ancestors and those citizens who are totally nomadic and completely isolated from the sedentary population. If one means by Bedouins the Saudi citizens who derive their livelihood from raising goats, sheep, and camels and follow the clouds (move to the parts of the country where it rained), then their number is perhaps less than 15 per cent of the population.

There is no doubt, however, that Saudi Bedouins are yesterday's majority who are becoming today's minority (Said, 1979). And their economic status has to be given greater attention in the coming years than it receives now.

(B) The state of the oil fields

In all of the Saudi oil fields gas and crude oil are 'trapped' under pressure in strata of porous rock; water surrounds the oil deposits. The permeability of the rock and the amount of the water pressure determine the amount of oil that can be recovered. Geologists tell us that the holes in the rock of Saudi oil reservoirs are large enough to allow oil to move freely toward the drilled wells.

The problem now (and it will be more so in the future) is the maintenance of the required pressure at a level that permits the recovery of the greatest possible amount of crude, given the state of technology.

Among the factors that govern the pressure is (obviously) the age of the field, the number of producing wells in a given area of a field, and the amount of oil pumped out from the entire field as well as that extracted from each well. That is, not only the total amount of oil extracted affects pressure, but also the rate at which extraction occurs and the number of wells used to accomplish the job.

Between mid-1970 and mid-1973, total Saudi oil output increased from 3.5 million b/d to 7.5 million b/d, an increase of more than 100 per cent, and most of this additional oil came from the Ghawar oil field. This field is the world's largest and stretches for nearly 250 kilometers, but still increasing its output in only 36 months by 4 million b/d eventually led to decreasing the pressure on its crude.

Since the early 1970s water had to be pumped in to replace the pressure lost as a result of accelerated oil production. Gas could also be pumped in to maintain pressure, but gas itself is a substitute for oil, and its value is rising with the increasing price of oil.

Of course, there is nothing wrong with pumping in water to maintain pressure at desirable levels, and pressure in every oil reservoir will eventually fall as the field ages and has to be replaced; the problem is that in Saudi Arabia sweet water is more scarce than oil. Thus, Aramco had to rely on sea water. And sea water creates problems that may be as serious as rapidly decreasing pressure.

Furthermore, injecting water, whether purified or not, creates new problems. As more oil is pumped out, the injected water seeps to the producing wells and they (the wells) start running 'wet'. Wet wells mean that water is mixing with oil, and in this case wells have to be blocked with concrete at the base or completely shut in.

Injecting sea water to elevate pressure creates other problems in addition to the possibility of wetting the wells; these are (a) the corrosion of pipelines and other equipment, (b) salt and other sea-water minerals may block the pores in the rock, resulting in a reduction of the amount of recoverable oil from the injected reservoirs.

What are the solutions to these serious problems? First, water has to be injected to prevent pressure from falling to levels that reduce the amount of recoverable oil. And in 1978 Aramco pumped in about 11.5 million b/d of saline aquifer water and treated sea water in the 'old' fields.

Secondly, the problems that water injection creates have also to be dealt with. One of the solutions, albeit an expensive one, is water desalination. In 1979 Aramco's sea-water treatment plant at Qurayyah in Eastern Ghawar started its operation. This plant, the world's largest when the whole system is completed, will produce 3.6 million b/d of treated sea water. The cost of the project may well exceed $1 billion. The problem of 'wet' wells that water injection generates had also to be dealt with. A plant to desalt crude is being built at Safaniyah, the main source of Saudi heavy crude. When it is finished in 1984, it will have a capacity of 2.1 million b/d of desalted crude.

Desalting plants will also be built at Berri, the premier producer of extra light crude, and in northern Ghawar.

It is estimated that 4 million b/d of crude oil will be desalted by the mid-1980s. The cost may reach $5 billion. (The source of information contained in this section is the *Saudi Business* issues of November 30, December 7, and Decembr 14, 1979.)

What we want to emphasize in this section is the following: the days when the operational costs of producing oil in Saudi Arabia were mainly the costs of drilling and operating wells are gone for ever. Substantial investment in sea-water treatment plants as well as in crude oil desalination plants has to be undertaken. And acceleration of oil outputs does not only reduce total reserves; it also reduces the amount of oil that otherwise would have been recoverable.

Since the Saudi economy is completely dependent on petroleum, it is rather an understatement to say that the delicate conditions of the old Saudi oil fields ought to be of concern to every Saudi citizen.

(C) Water

Water shortages in the Arab Peninsula are nothing new. What is really new about this in Saudi Arabia today is the high cost of producing water and the incredibly low prices people are charged for using it.

It is estimated that the Saudi urban centers consume about 1,200 million tons of water per year. The projected amount of water that Saudi cities and towns will consume in the year 2000 is 4 billion tons (*Saudi Business*, April 18, 1980: 10).

At the present time Saudi Arabia gets half of its drinking water from desalination plants; the other half comes from underground reservoirs—fossil water.

But fossil water is like oil, an unrenewable resource, and if Saudis continue consuming it in the future at the same rates as in the 1970s, none of it will be left for future generations.

Treating sea water is a very expensive alternative. The Saudi government has spent so far Rls 25 billion for water desalination plants and related projects; another Rls 7 billion will be spent soon. Furthermore, this Rls 32 billion will be only what is spent under the aegis of the Saline Water Conversion Corporation (SWCC) and does not include Aramco's desalters for oil operations or the Ministry of Defense plants or even the Ministry of Agriculture plants (*Saudi Business*, January 11, 1980: 16–19).

The vice governor of SWCC was quoted as estimating the 'real cost' of desalinated water to be around $2.00 a cubic meter. That is very expensive indeed. Yet it seems almost certain that the 'real cost' to Saudi Arabia of producing each additional cubic meter of water is much more than $2.00. Why? Because in Kuwait, where cost conditions should not be much different from what they are in Saudi Arabia, the government charges Kuwaiti consumers up to $4.00 for a cubic meter of desalinated water (*Saudi Business*, January 11, 1980: 17). And it is almost certain that if the real cost of a cubic meter of desalinated water in Kuwait is about $2.00 (as is being claimed to be the real cost in Saudi Arabia), then the Kuwaiti government will not ask consumers to pay $4.00 for it.

But since Saudi consumers are only charged Rls 0.5 (equivalent to $0.15), they are in effect encouraged to waste water *even* if each cubic meter costs only $2.00. If consumers pay for a commodity $0.15 while its cost of production is $2.00, they will behave as if the cost were actually $0.15—the amount of income they have to give up in order to acquire each additional cubic meter of water.

The Saudi government's motive in keeping the cost of water to consumers so low is rather obviously an attempt to keep the cost of living as low as possible. The problem is that this method of subsidizing consumers results in misallocation of resources.

Everybody would be better off if the real price of water was paid and the money then given back to consumers as a lump sum which they would be free to spend in any way they see fit. In this way consumers would receive subsidies which increase their real income without encouraging them to waste water. Then the relative prices of all the goods and services they consume as well as the level of their income will determine their water consumption. And given that the real price of water is going to be somewhere between 13 and 26 times its present level, consumers' income would have to reach many times its present level before they consume water at the rates at which they are consuming it now.

If by the year 2000 Saudis are going to consume 4 billion cubic meters of water, as currently projected, then the opportunity cost of consuming that much water would be at least $8 billion (4 × $2.00) in 1977 dollars. That would be almost one-third of the entire government budget.

It is rather apparent that something has to be done about conserving water. The most obvious and immediate remedy should be to increase the price of water to consumers.

But good news may eventually come from Saudi Arabia's Center for Science and Technology (SANCST) and from the University of Petroleum and Minerals' Research Institute. Both are engaged in applied desalination research, and scientists at SANCST as well as at UPM Research Institute think that there is a good chance that their research will lead to technological breakthroughs which will result in reducing the cost of 'harvesting' sweet water from the sea.

REFERENCES

Cole, D. P. (1975). *Nomads of the Nomads*, Aldine Publishing, Chicago, Illinois.
Said, A. H. (1979). *Saudi Arabia: The Transition From A Tribal Society to a Nation State*, Ph.D. Dissertation, University of Missouri, Columbia, Missouri.
SAMA, *Annual Reports*, 1970–1979.
Saudi Business, October 5, 1979.
Saudi Business, November 30, 1979.
Saudi Business, December 7, 1979.
Saudi Business, December 14, 1979.
Saudi Business, January 11, 1980.
Saudi Business, April 18, 1980.

CHAPTER 9

The World Oil Market and the Role of Saudi Arabia: A Summary

When I picked the Organization of Petroleum Exporting Countries as the subject of this study it did not occur to me for a single moment that OPEC may not be a cartel. In fact, the original tentative title was *OPEC's Future: Prospects and Problems*.

It turned out that, whatever OPEC's future is, it has very little importance anyway. Then, one would ask, how could we explain the sharp rise in the price of oil toward the end of 1973? In a more general way, the question we must ask is whether international oil prices reflect only the behavior of the individual oil countries acting as wealth maximizers, or do they reflect the collusive monopoly power that OPEC exercises?

It seems to me that the current oil price would have prevailed, independent of the presence or absence of OPEC, given that the individual oil-producing countries are the ones who determine rates of output and prices. I do not think that many will dispute my contention that the oil companies and the host countries have drastically different rates of discount. And different discount rates lead to different rates of output, and that leads to different prices.

Even the Executive Director of the Petroleum Industry Research Foundation, John H. Lichtblau, recognizes that the oil-producing countries' criteria in setting prices and rates of output are 'essentially different from those of the private companies'. He goes on to say,

> While the [companies] have an incentive to maximize production because of compelling short-term commercial imperatives, the [host countries] are more concerned with long-term national economic and social goals (Lichtblau, 1977: 509).

The difference in behavior between the companies and the host countries was mainly due to the companies' fear of nationalization, which led the companies to act like 'there is no tomorrow'. But the companies would have produced more oil even if they had faced no threat whatsoever to their property rights.

One reason is that the original concession agreements that gave the companies exclusive crude-oil ownership rights would have expired by the 1980s and 1990s. As a result, one would expect the companies to produce as much as possible before the expiration date.

The second reason is that most of the oil-producing countries do not have the same investment opportunities that the companies have.

Toward the end of 1973, everyone was aware that the price of oil rose sharply, but few knew (or fully appreciated) that a *de facto* nationalization of the companies' crude also took place. Due to the apparent suddenness and the magnitude of the price change, many economists have concluded that OPEC became an effective cartel that started to exercise its monopoly power. Furthermore, economists thought that, like other cartels, the oil cartel would soon crumble.

As was indicated in Chapter 3, economic theory does not teach us that cartels are always unstable. It teaches only that cartels could not continue to get high rates of monopoly profit. Once the members of a cartel agree to fix and (more importantly) abide by a price approaching monopoly levels, strong incentives are created for individual members to cheat.

In the case of the alleged OPEC cartel, Professor Adelman concluded,

> Unless the producing nations can set production quotas and, what is more important, obey them, they will inevitably chisel and bring prices down by selling incremental amounts at discount prices (Adelman, 1972: 258).

After explaining the incentive for cheating he concludes,

> The world oil cartel of the 1930s was eroded by this kind of competition, and so will be the new one [OPEC] in the 1970s.

As a consequence of the alleged OPEC cartel durability, Adelman changed his mind. His explanation, dealt with in Chapter 5, section 5.5, is that what accounts for the robustness of OPEC is the oil companies in their function as 'tax collectors' for the producing countries. That does not hold water because, analytically, there is no meaningful difference for an oil-producing country between reducing its oil price and reducing its marginal tax rate to increase its market share.

Other writers suggested that OPEC became more stable because the alleged cartel consists of sovereign states and therefore, unlike other cartels, did not worry about legal prosecution (Kalymon, 1975).

The conclusion is this: cartels, legal or illegal, that are able to earn monopoly profits for their members are inherently unstable. The main reason the oil price rose sharply since the end of 1973 is the change of property rights in crude oil.

To say that the cost of producing a barrel of oil in the Middle East is between twenty and thirty cents simply does not explain much. What matters to the oil-producing countries is the maximization of the discounted present value of their oil. Therefore, their decision as to how much oil to produce in each period is influenced by their discount rates and the expected future price, as well as by the current and future costs of exploitation. Given the limited investment opportunities in most of the producing countries, one could not say that their oil output rates are not consistent with wealth maximization behavior, aside from any effort on their part to keep their alleged 'cartel' together.

The OPEC members never engaged in demand prorationing or profit sharing. The economic and political differences between them would preclude the necessary cooperation to make their organization a truly effective cartel.

It seems to me (and one can only speculate in this regard) that the many able economists who were and are convinced that OPEC is a cartel just failed to recognize that the utility functions and constraints of the oil companies on the one hand and the host countries on the other are drastically different. As far as I know, no one paid any attention to the fact that, at the end of 1973, a shift of property rights had occurred.

From the late 1940s to the early 1970s, the real price of oil fell by about 65 per cent. The price of oil was then, and still is, being determined by the supply of and demand for oil. What has changed are the forces that determine the supply of crude oil.

In the 1950–72 period, the average annual gross additions to the free world's oil reserves were roughly 27 billion barrels. That meant the oil companies had a choice between increasing output, and thus reducing the prices, or keeping the prices high by leaving a lot of oil in the ground when their concessions' terms had expired. But any individual company could not have much of a choice because the oil market gradually became competitive, and any individual producer's efforts to prevent a price fall by output cutbacks would have only resulted in the reduction of its own revenues.

Since 1973, increases in reserves ceased to mean automatic increases in supply. As far as any oil-producing country is concerned, the time of exhaustion is not influenced by the threat of nationalization or the expiration of the concession terms. Thus the current oil price is not as arbitrary as is widely believed, and in fact reflects the true long-run supply costs more than the pre-1973 price did.

In other words, the price of oil is still being determined, as it has been for at least the last thirty years, by the demand for oil and by its long-run supply costs. The only thing that has changed since 1973 is the supply cost. What is a cost to a company may not be a cost for the host country, and *vice versa*.

When cartel models failed to explain the behavior of oil prices, price-leadership-by-a-dominant-firm models were employed to explain OPEC's behavior. So many writers confused 'cartelization' with price leadership, and as a result many readers were led to believe that 'cartel' and 'price leadership' meant the same thing, though in fact they do not.

In the period January 1974 through December 1977, Saudi Arabia seemed to be the oil price leader. And it was able to assume that role simply because it increased its rate of output whenever the other members of OPEC demanded higher prices. Its productive capacity in that period enabled it to increase its output by the amount required to prevent oil price rises.

If Saudi Arabia's efforts were directed toward preventing price decreases the job would have been harder, though it still would have been possible during that period. For whatever reason, the *real* price of oil in the period between January 1974 and May 1978 fell by about 40 per cent.

During the fall of 1978, Saudi Arabia increased its oil output by about 40 per cent to prevent the sharp rise in oil prices that followed the stoppage of Iranian oil exports. It did not succeed. And what happened since has shown that Saudi Arabia's ability to determine the world's oil price is less than what many people believe.

Today, there is no doubt in my mind that Saudi Arabia would have been better off without OPEC. The absence of OPEC may not change Saudi oil policy (which is determined by economic as well as noneconomic considerations) in regard to rates of oil output, but will allow it to change the 'world price' for its oil.

Many Saudi officials know that OPEC is neither monopolizing the world oil market nor in any other way keeping the price of oil higher than what it would have been without it, but still they cannot even imagine a world without OPEC. The reason: they think OPEC provides stability to the world oil market.

I do not think OPEC stabilizes or destabilizes the oil market, because as an organization it has no influence on the supply of oil or the demand for it. The argument that OPEC has a stabilizing effect is similar to the one that was advanced to defend the system of 'fixed' exchange rates for the world's currencies. Eventually, that system broke down because the 'fixed' rates of exchange (prices) were not what was dictated by the forces of supply and demand. In time this happened to the price that OPEC 'fixes' and for precisely the same reason—the OPEC price could not equate supply and demand.

Since the beginning of 1979, even the **appearance** of price unity has disappeared. In the first half of 1979, divergence between OPEC's price and prices in the spot market ranged from $20 to $10. Further, no oil producer was abiding by the 'OPEC price' except Saudi Arabia. The oil of every OPEC producer as well as Mexican and North Sea oil were all sold at a higher price than what the Saudis got for theirs.

The net result: Saudi Arabia is extracting more oil at rates which are not considered economically optimal, and furthermore it is selling it at lower prices than what every other oil-producing country is charging.

If the world wants Saudi Arabia to extract more oil, then it must make the oil revenues more useful by helping Saudis in their efforts toward expanding their economy's productive base.

The rate of growth of the Saudi economy in the last few years has been around 10 per cent. The most significant development occurred in the fields of communication and education.

Some of the national economic issues that are going to demand greater attention from Saudi policy makers in the future are income redistribution, declining pressure in some of the old oil fields, and the increasing scarcity of water.

REFERENCES

Adelman, M. A. (1972). *The World Petroleum Market*, Johns Hopkins University Press, Baltimore, Maryland.

Kalymon, B. A. (Spring 1975). 'Economic Incentives in OPEC Oil Pricing Policy', *Journal of Development Economics*, 337–362.

Lichtblau, J. H. (August 1977). 'Factors Shaping Future Petroleum Prices', *Oil and Gas Journal*, 509–11.

Bibliography

Adelman, M. A. 'The Changing Structure of Big International Oil' in *Oil, Divestiture, and National Security*, ed. by Frank N. Trager (New York: Crane, Russak, and Company, Inc., 1977), pp. 1–11.
Adelman, M. A. 'Is The Oil Shortage Real? Oil Companies as OPEC Tax Collectors', *Foreign Policy*, **8** (Winter 1972–73), 69–107.
Adelman, M. A. 'Politics, Economics, and World Oil', *The American Economic Review*, **64** (May 1974), 58–68.
Adelman, M. A. 'The World Oil Cartel', *Quarterly Review of Economics and Business*, **16** (April 1976), 3–11.
Adelman, M. A. *The World Petroleum Market*. Baltimore: Johns Hopkins University Press, 1972.
Adler, John. *Absorptive Capacity: The Concept and Its Determinants*. Washington, D.C.: Brookings Institution, 1965.
Akins, J. E. 'The Oil Crisis: This Time the Wolf is Here', *Foreign Affairs*, **51** (April 1973), 463–90.
Alchian, A., and Allen, W. *University Economics*, London: Prentice-Hall International, 1974.
Alnasrawi, A. 'Collective Bargaining Power in OPEC', *Journal of World Trade Law*, **7** (April 1973), 202–14.
Arab League (in Arabic). *The Efforts of the Arabs in Oil Affairs*. Cairo, April 1969.
Arabian American Oil Company (ARAMCO). *The Middle East Development*. Dhahran, Saudi Arabia, 1956.
Arabian American Oil Company (ARAMCO). *A Report of Operations to the Saudi Government*. Dhahran, Saudi Arabia, 1961.
Asfour, E. Y. *Saudi Arabia: Long-Term Projections of Supply and Demand for Agricultural Products*. Beirut: American University Beirut, 1965.
Bain, J. S. *Barriers to New Competition*. Cambridge, Mass.: Harvard University Press, 1956.
Baranyi, L., and Mills, J. C. *International Commodity Agreements*. Mexico: Centro De Estudios Monetarios Latino-americanos, 1963.
Barclay's Bank. *Economic Survey of Saudi Arabia*. August 5, 1977.
Becker, G. S. *Economic Theory*. New York: Knopf, 1971.
Becker, G. S. 'Crime and Punishment: An Economic Approach', *Journal of Political Economy*, **76** (April 1968), 164–207.
Bennathan, E., and Walters, A. A. 'Revenue Pooling and Cartels', *Oxford Economic Papers*, **21** (July 1969), 161–76.
Bradley, Paul G. 'Increasing Scarcity: The Case of Energy Resource', *The American Economic Review* (May 1973), 119–26.

Brems, H. 'On the Theory of Price Agreements', *Quarterly Journal of Economics*, **65** (May 1951), 252–63.
Central Bank of Libya. *Economic Bulletin* (July–August 1971), 186.
Central Department of Statistics in Saudi Arabia. *Statistical Yearbook, 1965 to 1976*. Riyadh.
Central Planning Organization in Saudi Arabia. *Development Plan*. Riyadh, 1978.
Chandler, G. 'Some Current Thoughts on the Oil Industry', *Petroleum Review*, **27** (December 1973), 6–12.
Chase Manhattan Bank. *Business in Brief*. Monthly Report No. 134, June 1977.
Chenery, H. B. 'Restructuring the World Economy', *Foreign Affairs*, **53**, 2 (January 1975), 242–263.
Chenery, H. B., and Strout, A. 'Foreign Assistance and Economic Development', *American Economic Review*, **56**, 4 (September 1966), 679–733.
Coale, A., and Hoover, E. M. *Population Growth and Economic Development in Low-Income Countries, A Case Study of India's Prospective*. Princeton: Princeton University Press, 1958.
Cole, D. P. *Nomads of the Nomads*. Chicago: Aldine Publishing, 1975.
de Chazeau, M. G., and Khan, H. E. *Integration and Competition in the Petroleum Industry*. New Haven: Yale University Press, 1969.
Dickson, H. R. P. *The Arab of the Desert*. London: Alden, 1949.
The Economist, London, England.
Edens, D. G., and Snavely, W. P. 'Planning for Economic Development in Saudi Arabia', *The Middle East Journal*, **24**, 1 (Winter 1970), 17–30.
El Mallakh, R. 'The Challenge of Affluence: Abu Dhabi', *The Middle East Journal*, **24**, 2 (Spring 1970), 135–46.
El Mallakh, R. 'The Economics of Rapid Growth: Libya', *The Middle East Journal*, **23**, 3 (Summer 1969), 308–20.
El Mallakh, R. *Economic Development and Regional Cooperation*. Chicago: The University of Chicago Press, 1968.
El Mallakh, R., and Kadhim, M. *Personal Communication with CACI*. Boulder: University of Colorado.
Emirate of Abu Dhabi. *Statistical Abstract III and Yearbook 1974*. Abu Dhabi, U.A.E.: Department of Planning, 1974.
Energy Research and Development Administration (ERDA). *A National Plan for Energy Research, Development, and Demonstration: Creating Energy Choices for the Future*, Vols. 1, 2. Washington, D.C.: Energy Research and Development Administration.
Esmara, H. 'Regional Income Disparities', *Bulletin of Indonesian Economic Studies*, **11**, 1 (March), 41–57.
Faisal, Prince. *Ten Point Program*. Riyadh: Saudi Arabia, November 6, 1962.
Fallon, N. 'Oil Reserves Estimates Due for Revision', *Middle East Economic Digest* (May 10, 1974), 528–29.
Farley, R. *Planning for Development in Libya: The Exceptional Economy in the Developing World*. New York: Praeger, 1971.
Federal Energy Administration (FEA). *The Relationship of Oil Companies and Foreign Governments*. Washington, D.C.: Federal Energy Administration, Office of International Energy Affairs, 1975.
Federal Energy Administration (FEA). *Oil: Possible Levels of Future Production*. Federal Energy Administration, Project Independence Blueprint, Final Task Force Report, November 1974.
Federal Republic of Nigeria. *Third National Development Plan 1975–80*, vols. 1, 2. Lagos, Nigeria: Federal Ministry of Economic Development.
Federal Republic of Nigeria. *Second National Development Plan*. Lagos, Nigeria: Federal Ministry of Economic Development.

Felliner, W. *Competition Among the Few*. New York: Frank Cass and Company, Ltd., 1965.
Fenelon, K. G. *The United Arab Emirates: An Economic and Social Survey*. London: Longman, 1973.
Fenelon, K. G. *The Trucial States: A Brief Economic Survey*. Beirut: Kyayats, 1969.
Fesharaki, F. *Development of the Iranian Oil Industry*. New York: Praeger, 1976.
Field, P. 'Saudi Arabia: How the $142,000 Million Will be Spent', *Middle East Economic Digest* (August 22, 1975), 5–6, 28–32.
The 5th Plan, Revised and Summarized Version Supplement No. 228. Tehran, March 1975, n.p.
Firoozi, F. 'The Iranian Budgets: 1964–1970', *International Journal of Middle East Studies*, **5** (1974), 328–43.
First, R. *Libya: The Elusive Revolution*. London: Penguin Books, 1974.
First National City Bank. 'Why OPEC's Rocket Will Lose Its Thrust'. New York: First National City Bank, Economic Department, *Monthly Economic Letter* (June 1975), 11–15.
Fischer, G., and Muzaffar, A. M. 'Some Basic Characteristics of the Labor Force in Bahrain, Qatar, United Arab Emirates, and Oman'. Paper delivered to Conference on Manpower in the Arabian Gulf, Manama, Bahrain, February 1975.
Fleming, J. J. 'External Economics and the Doctrine of Balance Growth' *Economic Journal*, **65**, 258 (June 1955), 241–56.
Fog, B. 'How Are Cartel Prices Determined?' *Journal of Industrial Economies*, **5** (November 1956), 16–24.
Frank, H. *Crude Oil Prices in the Middle East: A Study in Oligopolistic Price Behavior*. New York: Praeger, 1966.
Frankel, P. H. 'The Oil Industry and Professor Adelman', *Petroleum Review*, **27** (September 1973), 347–49.
Frankel, P. H., and Newton, W. L. 'Comparative Evaluation of Crude Oils', *Journal of the Institute of Petroleum*, **56** (January 1970).
Geer, T. *An Oligopoly, The World Coffee Economy and Stabilization Schemes*. New York: Dunellen Publishing Company, Inc., 1971.
Gordon, R. L. 'A Reinterpretation of the Pure Theory of Exhaustion', *Journal of Political Economy*, **75** (June 1967), 274–86.
Gray, L. C. 'Rent Under the Assumption of Exhaustibility', *Quarterly Journal of Economics*, **28** (May 1914), 466–89.
Heravi, M. (ed.). *Concise Encyclopedia of the Middle East*. Washington, D.C.: Public Affairs Press, 1973.
Herfindahl, O. C. 'Depletion and Economic Theory' in *Extractive Resource and Taxation*, ed. by M. Gaffney. Madison, Wisconsin: University of Wisconsin Press, 1967.
Herfindahl, O. C., and Knesse, A. *Economic Theory of Natural Resources*. Columbus, Ohio: Charles E. Merill Publishing Company, 1974.
Hexner, E. *International Cartels*. Durham, N.C.: University of North Carolina Press, 1946.
Hitti, S. H., and Abed, A. T. 'The Economy and Finance of Saudi Arabia', *IMF Staff Papers*, **21**, 2 (July 1974), 247–307.
Hirschleifer, J. 'Investment Decision Under Uncertainty: Applications of the State Preference Approach', *Quarterly Journal of Economics*, **80**, 2 (May 1966), 252–77.
Horvat, B. 'The Optimum Rate of Investment', *Economic Journal*, **68** (December 1958), 747–67.

Hotelling, Harold. 'The Economics of Exhaustible Resources', *Journal of Political Economy*, **39** (April 1931), 137–76.
Iskandar, M. *The Arab Oil Question*. Beirut: Middle East Economic Consultants, 1974.
Iskandar, M. 'Economic Development Plans in Oil Exporting Countries and Their Implications for Oil Production Targets' in Z. Mikdashi, S. Cleland and I. Seymour (eds.), *Continuity and Change in the World Oil Industry*. Beirut: The Middle East Research and Publishing Center, 1970.
Issawi, C. 'Iran's Economic Upsurge', *The Middle East Journal*, **21**, 4 (Autumn 1967), 447–61.
Issawi, C., and Yeganeh, M. *The Economics of Middle Eastern Oil*. New York: Praeger Publishers, 1962.
Jacoby, N. H. *Multinational Oil*. New York: Macmillan Company, 1974.
Kalymon, B. A. 'Economic Incentives in OPEC Oil Pricing Policy', *Journal of Development Economics*, **2** (1975), 337–62.
Knauerhase, R. *The Saudi Arabian Economy*. New York: Praeger Publishers, 1975.
Knorr, K. E. *Tin Under Control*. California: Stanford University Press, 1945.
Kubbah, A. *OPEC: Past and Present*. Vienna: Petro-Economic Research Center, 1974, p. 61.
Leeman, W. A. *The Price of Middle East Oil*. Ithaca: Cornell University Press, 1962.
Levy, W. J. 'Oil Power', *Foreign Affairs*, **49** (July 1971), 752–68.
Lichtblau, J. H. 'Factors Shaping Future Petroleum Prices', *Oil and Gas Journal*, **65** (August 1977), 509–11.
Machlup, F. *The Economics of Sellers Competition*. Baltimore: The Johns Hopkins Press, 1952.
Machlup, F. *The Political Economy of Monopoly*. Baltimore: The Johns Hopkins Press, 1952.
McDoland, S. L. *Petroleum Conservation in the United States: An Economic Analysis*. Baltimore: The Johns Hopkins Press, 1971.
McKee, J. W. 'The Political Economy of World Petroleum', *American Economic Review*, **64** (May 1974), 51–53.
Mead, W. J. 'An Economic Analysis of Crude Oil Prices Behavior in the 1970s', *The Journal of Energy and Development*, **4** (Spring 1979), 212–228.
Medvin, N. *The Energy Cartel*. New York: Random House, Inc., 1974.
Mikdashi, Z. 'Collusion Could Work', *Foreign Policy*, **12** (1974), 57–68.
Mikdashi, Z. *The Community of Oil Exporting Countries*. New York: Cornell University Press, 1972.
Mikdashi, Z. *A Financial Analysis of the Middle Eastern Oil Concessions, 1901–1965*. New York: Praeger, 1966.
Mikdashi, Z. *The International Politics of Natural Resources*. New York: Cornell University Press, 1976.
Mikdashi, Z. 'The OPEC Process', *Daedalus*, **104** (Fall 1975), 104–28.
Middle East Economic Digest. 'Rising Demand for Oil Met in 1970', *Middle East Economic Digest*, 15 (January 8, 1971), 33.
Middle East Economic Digest. 'Oil Talks Miss Deadline', *MEED*, **15** (February 5, 1971), 140.
Middle East Economic Digest. 'OPEC Threatens Oil Embargo', *MEED*, **15** (February 12, 1971), 165.
Middle East Economic Digest. 'Gulf Oil Producers Win Price Struggle', *MEED*, **15** (February 1971), 183.
Middle East Economic Digest. 'Five-Year Oil Pact Signed in Tripoli', *MEED*, **15** (April 9, 1971), 371.

Middle East Economic Digest. 'Oil Prices Agreed: Participation Talks Now', *MEED*, **16** (January 28, 1972), 97–8.
Middle East Economic Digest. 'ME Oil Output as US Crude Imports Rise', *MEED*, **17** (September 14, 1973), 1059.
Middle East Economic Digest. 'Unilateral Oil Price Increase', *MEED*, **17** (October 19, 1973), 1214.
Middle East Economic Digest. 'Auctioned Oil Fetches Record Prices', *MEED*, **17** (December 14, 1973), 1452.
Middle East Economic Digest. 'Khene Urges Cut in Energy Consumption', *MEED*, **18** (April 26, 1974), 474.
Middle East Economic Digest. 'What The New Oil Price Means', *MEED*, **18** (December 20, 1974), 1560–61.
Middle East Economic Digest. 'Oil-Exporters Rise to the Development Challenge', *MEED*, **19** (January 17, 1975), 4–6.
Middle East Economic Digest. 'Middle East Accounts for 42.5% of World Oil Output', *MEED*, **19** (January 24, 1975), 6.
Middle East Economic Digest. 'Estimates of OPEC Funds Revised', *MEED*, **19** (January 31, 1975), 5.
Middle East Economic Digest. 'No OPEC Move on Prices Likely', *MEED*, **19** (March 7, 1975), 7.
Middle East Economic Digest. 'OPEC Output Down 15 Per cent in March', *MEED*, **19** (May 19, 1975), 8.
Middle East Economic Digest. 'Nazer Explains Saudi Five Years Plan', *MEED*, **19** (June 6, 1975), 8.
Middle East Economic Survey, Nicosia, Cyprus.
Monthly Energy Review, US Department of Energy, Washington, D.C.
Moorsteen, R. 'OPEC Can Wait—We Cannot', *Foreign Policy*, **18** (Spring 1975), 1–12.
Nordhaus, W. D. 'The Allocation of Energy Resources', *Brookings Papers on Economic Activity*, **3** (1973), 529–70.
Okita, S. 'Natural Resource Dependency and Japanese Foreign Policy', *Foreign Affairs*, **52** (July 1974), 719–25.
Orr. D., and MacAvey, P. W. 'Price Strategies to Promote Cartel Stability', *Economica*, **32** (May 1965), 186–97.
Osborn, D. K. 'Cartel Problems', *American Economic Review*, **66** (December 1976), 835–44.
Patinkin, D. 'Multiple-Plant Firms, Cartels, and Imperfect Competition', *Quarterly Journal of Economics* (February 1947), 173–232.
Penrose, E. T. 'The Development of Crisis', *Daedalus*, 104 (Fall 1975), 39–59.
Penrose, E. T. *The Petroleum Industry*. Hearings before the Subcommittee on Antitrust and Monopoly of the Committee on the Judiciary, U.S. Senate, 91st Congress, 1st Session, Part I, pp. 432–33.
Petroleum Economist, London, England.
Petroleum Intelligence Weekly, New York, N.Y.
Prest, A. R., and Turvey, R. 'Cost–Benefit Analysis: A Survey', *Economic Journal*, **75** (December 1965), 683–735.
Quirk, J. R. *Intermediate Microeconomics*. Chicago: Science Research Associates, Inc., 1976.
Radetzki, M. *International Commodity Market Arrangements*. London: C. Hurst and Company, 1970.
Rifai, T. *The Pricing of Crude Oil*. New York: Praeger Publishers, 1974.
Robinson, J. *The Economics of Imperfect Competition*. London: Macmillan and Company Ltd., 1965.
Rohani, F. *A History of OPEC*. New York: Praeger Publishers, 1971.

Ronall, J. O. 'Banking Regulations in Saudi Arabia', *Middle East Journal*, **21**, 3 (Summer 1967), 399–402.
Rugh, W. 'Emergence of a New Middle Class in Saudi Arabia', *Middle East Journal*, **27**, 1 (Winter 1973), 7–20.
Russell, M., and Bohi, D. R. *U.S. Energy Policy: Alternative for Security*. Baltimore: The Johns Hopkins Press, 1975.
Safer, A. E., and Mills, A. M. 'The Outlook for World Oil: Prices and Petrodollars', *Business Economics* (September 1975), 21–31.
Said, A. *Saudi Arabia: The Transition From A Tribal Society to A Nation State*, Ph.D. Dissertation, University of Missouri, Columbia, Missouri, 1979.
Sampson, A. *The Seven Sisters*. New York: The Viking Press, 1975.
Saudi Arabian Monetary Agency (SAMA). *Annual Reports: 1970–1979*. Jeddah: Saudi Arabia.
Saudi Business, Jeddah, Saudi Arabia.
Sayegh, K. S. *Oil and Arab Regional Development*. New York: Praeger Publishers, 1968.
Scherer, F. M. *Industrial Market Structure and Economic Performance*. Chicago: Rand McNally Company, 1970.
Shepherd, W. G. *Market Power and Economic Welfare*. New York: Random House, 1970.
Stigler, G. J. *The Organization of Industry*. Homewood, Illinois: Richard D. Irwin, Inc., 1968.
Stigler, G. J. *The Theory of Price*. New York: Macmillan Company, 1966.
Stocking, G. W., and Watkins, M. W. *Cartels or Competition*. New York: The Twentieth Century Fund, 1948.
Stocking, G. W., and Watkins, M. W. *Middle East Oil*. Kingsport, Tenn.: Vanderbilt University Press, 1970.
Streeten, P., and Elson, D. *Diversification and Development: The Case of Coffee*. New York: Prager Publishers, 1971.
Telser, L. G. *Competition, Collusion and Game Theory*. Chicago: Aldine, Atherton, 1972.
Tippetts, C. C., and Livermore, S. *Business Organization and Control*. New York: D. Van Nostrand Company, 1932.
Weinstein, M. C., and Zeekauser, R. J. 'The Optimal Consumption of Depletable Natural Resources', *Quarterly Journal of Economics*, **89** (August 1975), 371–92.

Index

Abd-al Aziz, King 81–82
Absorptive capacity 61–63
   and the Saudi economy 63
Adelman, M. A. 5, 11, 13, 52, 96
Airports 74
Akins, James 11
Alchian, A. 20
Allen, W. 20
Al Qusaibi, Ghazi 82
Arab–Israeli conflict and oil 10
Aramco 91, 93
Arms-length deals 14
Assets, financial 62
   near money 69

Basing point 8
Becker, Gary 2
Bedouins 81–82
   the economic condition of 90–91
   and income distribution 90
   the number of 90–91
Berri oil field 92

Capital market, the absence of in Saudi Arabia 76
Cartels, the theory of 19–20, 21–25
Changes, structural 8–9
Colluding, the difficulty of 19–20, 25–26
Competition and monopoly 17–19
Concessions, terms of 4
Congestion 74
Consortium, the oasis 10
Council of Ministers 82, 83

Decision making in Saudi Arabia 82–83
Demand pro-rationing and OPEC 27
Desalination 92, 93, 94
Distribution, income 89–90

Economic development in Saudi Arabia 84–89
Economic issues 89–94
Education 86
Embargo 7, 14
Exhaustion, definition of 33
Extraction costs 35
EXXON and Libya 9–10

Fields, oil 91–92
Finance, the Ministry of 83–84
Financial assets 62
Foreign investment 66
   the risk of 66–70
Fossil water 93
Frankel, P. H. 26

Gas, the gathering of 89
Getty and the Algerian government 9
Ghawar oil fields 91, 92, 93
Gross Domestic Product 85

Housing 86

Ibn Saud 81–82
Import quota 5
Income distribution 89–90
Industrialization 87–88
Inflation, and costs 74
   'cost push' 74–75
   defined 73
   the nature of 73–75
   and the Saudi economy 63, 66, 73–75, 85
Investment, domestic 65–66
   foreign 66–70
   private Saudi 77–78
Iranian Revolution, and spot prices (The) 56–57
   temporary effect of 54

Issues, economic 89–94

Land, the price of 75–77
Law of Diminishing Returns (The) 63
Libya and OPEC 9–13
Lichtblau, J. H. 95

Mead, W. J. 49
Mikdashi, Z 26
Millionaires, Saudi 90
Money supply 63
Monopoly and competition 17–19

Nazir, Hisham 82

OAPEC, the formation of 6, 7
Oasis consortium (The) 10
Oil companies, 52
  the 'independents' 7–8, 47
  the profits of 58
  state-owned 8, 47
  the seven sisters 7–8
  as tax collectors (The) 52
Oil fields, the state of 91–92
Oil revenues 63, 85
OPEC,
  the cause of the creation of 5
  and collusion 20–21
  and demand pro-rationing 27
  the formation of 6
  the future of 95
  as a joint sales agency 26–27
  and Libya 9–13
  and output reductions 44–46
  and price fixing 45
  and production quotas 27
  and Saudi Arabia 49–50, 71–72
Oxy and Libya 9–10

Participation 12
Penrose, E. T. 8, 43
Petroleum, Ministry of 83
Petromin 88, 89
Ports 74, 85
Posted price 4, 5
Pressure and rate of oil output 91
Price leadership, by a dominant
    firm 28–32
  and Saudi Arabia 29, 49–51

Prices, the explosion of 56–58
  nominal 41
  the quadrupling of 13–15
  real 41
Price searcher 18
Price takers 18
Productive capacity 66, 68
Profits and the oil companies 58
Profit sharing 4, 5
Property rights, and the discount
    rate 37–39
  the change of 14–15, 43–44
  the uncertainty of 37–39

Real Estate Development Fund 86
Religion in Saudi Arabia 81–82
Rent controls 79–80
Roads 86

SABIC 88, 89
SAMA 63, 73
Saudi Arabia and OPEC 71–72
Saudi millionaires 90
Scarcity 76
Seven Sisters, the role of (The) 7, 8, 9
Sofaniyah oil field 92
Solow, Robert 3
Spot market 54–55
Spot prices 56–58
State-owned companies 8, 47
Stigler, George 2, 17
Structural changes 8–9

Teheran agreement (The) 11, 12
Telephone service 86
Tripoli agreement (The) 11, 12

User costs 35

Water 93–94
  the cost of producing 93–94
  desalination of 92, 93, 94
  injection of 92

Yamani, Ahmad Zaki 82

Zakat 90